BUILDING INSPECTION MANUAL

BUILDING INSPECTION MANUAL

A GUIDE FOR BUILDING PROFESSIONALS FOR MAINTENANCE, SAFETY, AND ASSESSMENT

KARL F. SCHMID

MOMENTUM PRESS

MOMENTUM PRESS, LLC, NEW YORK

Building Inspection Manual: A Guide for Building Professionals for Maintenance, Safety, and Assessment
Copyright © Momentum Press®, LLC, 2014.

First published by Momentum Press®, LLC
222 East 46th Street, New York, NY 10017
www.momentumpress.net

ISBN-13: 978-1-60650-615-8 (hard back, case bound)
ISBN-10: 1-60650-615-3 (hard back, case bound)
ISBN-13: 978-1-60650-616-5 (e-book)
ISBN-10: 1-60650-616-1 (e-book)

DOI: 10.5643/9781606506165

Cover design by Jonathan Pennell
Interior design by Exeter Premedia Services Private Ltd.,
Chennai, India

10 9 8 7 6 5 4 3 2 1

Printed in the United States of America

CONTENTS

PREFACE

It all begins with inspections. Everyone who is involved with buildings, or even a single building, whether as an owner, designer, constructor, operator, maintainer, buyer, or seller, will eventually have to conduct or have conducted facilities[1] inspections.

Throughout the United States, cities and counties have a building department. Sometimes they are named the Department of Buildings, sometimes Department of Safety and Buildings, and there may be other names too. But they have this in common: (1) The plan reviewers ensure code compliance and (2) the inspectors ensure that the construction is in accordance with the approved plans. Both plan reviewers and inspectors are usually knowledgeable of the local building codes (though rare, it's a problem when they are not), so, while at times these may seem difficult, these should be viewed as part of the project team. There are checklists that will help you to prepare for the inevitable building department plan reviews and inspections.[2]

Before buying or leasing a facility, a potential owner/lessee needs to know about its structural and mechanical/electrical systems and whether it can serve its intended use. Questions such as these (for sure an incomplete list of questions) need to be asked: has maintenance been deferred; how well does the facility stack up to Leadership in Energy and Environmental Design (LEED) standards, that is, did the facility's design and prior maintenance take energy efficiency and sustainability into account; what will it cost to bring the buildings up to current codes and standards; is the facility in compliance with the Americans with Disabilities Act (ADA); are there any hazards that require immediate attention; and, how should deficiencies be prioritized? There are checklists that are intended to help you prepare and conduct your due diligence/capital project planning inspections.

Maintenance and capital improvement inspections from which maintenance and capital improvement plans are developed should be conducted on a systematic and regular basis. These inspections identify and quantify the condition and functional performance of facilities and lead directly to annual and long-range plans and cost estimates for correcting deficiencies. Regular inspections provide a database that can be used as a baseline for future condition inspections. When properly documented inspections identify wear patterns and poorly functioning equipment, develop energy consumption and other trends, identify regular maintenance requirements, and provide planning tools for annual facilities

---

[1] Facility/facilities: Here the terms "facility" and "facilities" are sometimes used rather than "building" and "buildings." Facility is intended to suggest a building and its surrounding grounds, utilities, and systems. Similarly, facilities is intended to suggest more than one building along with their grounds, utilities, and grounds, for example, a campus.

[2] This is not to say that this book will meet every need every time. When it doesn't fully meet your needs, I hope that it serves as a tutorial that guides your research and preparation for your next inspection.

maintenance, repair, and long-term capital renewal budgets. Such systematic and regular inspections are vital to extending the useful life of the facilities by identifying potential problems before they occur; identifying out-of-sync and, thus, wasteful energy using equipment; and reducing disruptions due to equipment downtime. Over time, financial reports inform you about the health of an organization, so too over time do inspection reports inform you of the health of the organization's facilities.

The first two sections of this book fall under the heading New Building Construction: New Light Construction Inspections[3] and High-Rise Building Construction. The checklists assume on-going new construction that will be inspected by city and county building inspectors. However, the mission of all city and county building departments is to ensure the public safety, usually through enforcement of the building codes, you too are preparing for the long-term safe occupancy of the facility being inspected. The inspectors may seem like a pain-in-the-neck, but when all is said and done, they are serving your best interests.

The next two major sections fall under the heading Due Diligence and Existing Building Capital Project Planning Inspections: House/Light Construction and Larger Facilities. The checklists assume existing facilities and that the information gathered is for someone who is planning to purchase or lease the facility or to repair and maintain it.

At the end of this book, in Appendix A, I discuss briefly how New York City's School Construction Authority approaches the issue of keeping up with maintaining approximately 1,200 schools throughout the city of New York. Appendix B provides a list of construction industry resource organizations.

Each section is laid out using a different format. This was done to suggest that there is no one right way to approach your inspection. How you go about conducting your inspection and reporting is totally your option. Most (but not all) of the checklists ask questions that can be answered with "Yes," "No," or "Not Applicable (NA)" or "Satisfactory," "Unsatisfactory," or "Not Applicable." Yes or No answer generally requires no immediate further action, but a No answer indicates a problem. (There are places where it was not practical to hold to this format. I believe they are self-evident). Where the answer is No, you will have to give the item some further thought and make a comment in your inspection report; a cost estimate may be required. You can edit the questions into any form that you wish to meet your specific needs. Also, while the questions are asked under one category of inspection, for example, New Light Construction Inspections, they are transferable; I have assumed that you, the reader, are not a construction novice and, thus, will be able to pick and choose from all of the sections to meet your specific needs of the day. Saying the same thing in a different way, I have tried not to repeat myself, assuming that as you plan for an inspection for, say, a high-rise building, you will have looked at all the checklists and, where applicable, selected inspection items from them. And, beware, the questions do not cover every possible inspection item. Hopefully, they cover enough items for you to conduct a fully satisfactory inspection.

Lastly, my first time through this book, I attempted to identify by title the inspector for whom the questions were tailored, for example, a Department of Buildings' inspector. This became awkward. Contractors, architects, engineers, and owners must also prepare for these inspections, so the questions apply equally to them. In the end, I decided to simply use words such as, "as you prepare for your inspection," assuming that whoever is reading this book will be using the checklists for the purposes intended; it will be your inspection.

---

[3] As used in this section light construction refers generally to standard rafters, wood stud walls, and floor and ceiling joists. Examples of light construction are residential buildings, small office buildings, mini-malls, and the like.

ABOUT THE AUTHOR

Karl F. Schmid, PE, LEED AP, is a construction consultant based in Los Angeles, California. He holds a BCE from the City College of New York, an MS degree in civil engineering from Stanford, and an MBA from Cornell University. He has held positions with the U.S. Army Corps of Engineers, Cornell University, City University of New York, and the City of New York's Department of Buildings. Besides this book, Mr. Schmid has authored *Construction Crew Supervision*, *Construction Estimating*, and the *Concise Encyclopedia of Construction Terms and Phrases*, also published by Momentum Press.

PART 1
NEW BUILDING CONSTRUCTION INSPECTIONS

Former Speaker of the House Tip O'Neill coined the phrase, "All politics is local." It succinctly captures the principle that a politician's success is directly tied to his ability to understand and influence the issues of his constituents. Similarly, "all building codes are local." You may have an image of a group of architects, engineers, and contractors gathered around a table to write the building code for their community and that when completed represents the finest technical document possible for the community. But that's not how it normally works. As the code is being written, now usually beginning with a model code such as the International Residential or Building Code, various interest groups get involved. For example, plumbers and firefighters may argue against plastic piping, carpenters against metal studs, landlords against retroactively installing certain safety devices, and on and on. Some will prevail. Therefore, it is virtually impossible to write inspection checklists that are universally applicable. Furthermore, each community decides what inspections it wants conducted. This is usually a function of bad experiences and available funds, so no two communities conduct the same kind and number of inspections. And then, each inspector inspects based on his or her experience and knowledge.

Nevertheless, there are commonalities. For example, in almost every place in the United States, before any new building can be built, a local government (building department) plan reviewer would have checked the plans to ensure that they comply with all applicable building codes and then issued a building permit. Also, before any permit was issued, clearances probably had to have been obtained from other agencies such as Zoning, Planning, Fire Department, Coastal Zone Commission, and Historical Area Commission. And, in some areas you may also have been required to submit along with your request for a plan review a geotechnical report that was prepared by a licensed professional engineer. For example, in Los Angeles, California, where there are steep slopes that are both earthquake and fire prone, geotechnical reports are a common requirement of the package that is submitted for a plan review.

Whether as an owner, architect, engineer, or contractor, you prompt the plan review when you apply for a permit. Come prepared: plans should be clear, legible, and correctly show the work to be done. Usually, for light construction you will be required to submit a plot plan of the building on the lot; floor plans of the proposed work; a framing plan that shows the size and spacing of all structural members; a foundation plan that includes floor-framing details; elevations of new buildings or additions and construction details for stairways, chimneys, and similar improvements; construction details that indicate the size and direction of rafters, joists and studs; and energy conservation details, such as insulation. If you are intending to construct a high-rise building, the principles just stated remain more-or-less the same, but the input that you will be expected to provide will be multiples more, escalated very much in proportion to the size of your building and this goes for the other agencies such as zoning too.

Your permit will be issued based on the information that you provide, and then construction inspectors will follow-up to ensure that the work is in accordance with these plans and specifications. Sometimes a single local building inspector would conduct the on-site inspections to ensure compliance with the codes and plans. In larger communities inspectors with special knowledge, for example, electrical and plumbing will conduct a trade-only inspection. In the case of high-rise construction you can count on several, perhaps even numerous, inspectors visiting your site. In either case, that is, light or high-rise construction, besides building department inspectors you may be inspected by people from other agencies: for example, fire department, water, sewer and power, public works, OSHA, ADA.

Before a new building or addition can be occupied, or a change of occupancy can occur and sometimes following alterations and repairs above a certain size or dollar amount the owner needs to be issued a Certificate of Occupancy (CofO). Before a CofO is issued, two things must usually occur: A building inspector must perform a final inspection to ensure a project's conformance with approved plans and all applicable code requirements. Other departments such as engineering, city planning, and fire department may also have to certify that they have approved the work that falls under their jurisdiction. Once all the inspections are completed, you will have to apply for and receive a CofO before the building can be occupied.

1
HOUSE/LIGHT CONSTRUCTION

The checklists in this section are intended to assist you in preparing for building department inspections. They do not address all possible conditions, include all possible inspection items, or attempt to duplicate complete building codes. Hopefully, they do provide you with an understanding of what to expect as to the breadth and detail of the inspections that building department inspectors conduct, so that you can prepare for your unique situation. If you are an owner, the checklist should help you in assessing your contractor's work; they may also prompt you to get more deeply involved in your project.

The checklists have been structured to the extent possible as questions for which you are seeking Yes or Not Applicable (NA) as an answer. There may be many NAs as there are questions that apply to, say, flat roofs and sloped roofs while you probably have one or the other but not both. You can expect the building department to conduct the following inspections as construction progresses;[1] usually multiple visits are made for each category as the construction progresses.

- Temporary power
- Site work
- Basement/cellar walls, footings, and slabs
- Electrical rough-in
- Mechanical rough-in
- Plumbing rough-in

---

[1] The inspections listed are common but not inviolate. For example, the City of Lubbock, TX, requires the following inspections: (1) Footing; (2) plumbing rough; (3) return air (if applicable); (4) slab inspection; (5) plumbing top out; (6) mechanical rough duct; (7) electrical rough-in; (8) electrical underground service (if applicable); (9) flat work; (10) sewer and service lines (water and gas); (11) electrical final; (12) gas final (if applicable); (13) plumbing final; (14) mechanical final; and (15) building final (before the house/building is occupied).

- Framing
- Moisture, thermal protection, and insulation
- Final inspection(s)

Fees for these inspections may be included in the building permit fee, or there may be separate fees for each inspection. Contractors are usually required to schedule these inspections and work cannot progress beyond a certain point until an inspection is completed and approved, for example, walls cannot be closed-in until electrical rough-in has been approved. Having to wait around for an inspector to arrive can be upsetting, especially if workers are standing around, so it is important to schedule the inspection ahead of time, but not so far ahead of time that the inspector arrives before all necessary work is completed. Inspectors do not do physical work, so many people view them as useless government interlopers. Understand that the inspectors' role is to protect public safety. They do this primarily by enforcing all applicable building codes.

At the risk of raising a sore subject, it is easy to see how corruption creeps into the process. Building department inspectors have lots of power, and unscrupulous inspectors can hold up a job by not arriving on time and by unfairly disapproving work. Contractors on the other hand want to keep their jobs moving, so a bribe may seem like the easy way. My advice to both parties is *don't*. I have seen both inspectors and contractors go to jail. In the case of the inspectors, besides jail time they lost their pensions. In the case of the contractors, besides jail time they lost their businesses and savings. There are places to report an unauthorized "hand-out": police, FBI, state licensing board, or the head of the building department; you'll have to find your way. But don't let yourself get involved in paying off an inspector. End of lecture!

Before launching into the checklists, be aware that most jurisdictions require that your project's address and all permits be posted so that they can be easily seen from the approach road. Also, that approved plans and specifications for all engineered systems are on site and accessible to an inspector. Proof of all electrical, mechanical, fire sprinkler, and plumbing rough-in inspections and prior building inspections should also be maintained on-site.

In the checklists that follow an effort has been made to avoid specific numbers such as dimensions, distances, and times. However, where given, you should consider them as examples. Check to ensure that they agree with the applicable local codes.

TEMPORARY POWER

Whether from a permanent or temporary installation, a shock or blast can be equally deadly. And either installation can ignite a fire if the conductors overheat or if faulty wiring produces an arc. Therefore, essentially the same rules that pertain to workmanship, conductor ampacity[2] limits, and overcurrent protection apply to temporary power as to permanent power.

The National Electric Code (NEC), Art. 590, does, however, allow some modifications in certain wiring methods and materials. Temporary wiring is allowed

---

[2] Ampacity: The amount of current a conductor can carry without exceeding its specified temperature, in amperes.

only for construction, remodeling, maintenance, repair, or demolition of buildings, structures, or equipment, or similar activities and for emergencies, tests, experiments, and developmental work. Temporary installations have expiration dates at which time temporary wiring must be removed immediately upon completion of construction or the purpose for which it was installed; in some cases there is a specific time limit for the use of a temporary installation, for example, a holiday display.

Because removal is inherent in temporary power installations non-metallic (NM) type cable is allowed without a height limitation, whereas a permanent installation would require a raceway or metal-sheathed cable-type wiring. Thus, temporary installations do provide some savings on installation costs and are easier to remove when compared to more permanent installations.

When a temporary power installation is required, it is important that you learn who the authorities having jurisdiction (AHJ)[3] are. Besides the local building department inspector and utility company inspector, there may be other AHJs over your temporary installations. These may include: OSHA[4]; the general contractor, site owner, and occupant; insurers, corporate safety managers, and VPs of operations; and local fire marshals, city engineers, and other permit-granting authority. Temporary wiring methods are acceptable only if all the AHJs approve them.

The NEC speaks to the electrical hazards of temporary installations, but there are yet other dangers; you as the inspector should be always alert when in and around a temporary electrical installation. Tripping hazards, spill hazards, and the hazards of things falling on people are common construction site hazards.

- Has a utility company inspector inspected and approved underground electrical work? Notes: (1) It is common for a utility company to inspect and tag underground work prior to a local government inspector coming on site. (2) Usually the utility energizes the temporary installation.
- Are all underground service conduits sealed around entrance conductors?
- Are outdoor receptacles weatherproofed regardless of whether the attachment plug cap is inserted? Note: "While-in-use" receptacle covers are generally required.
- Is there adequate working clearance in front of all panels? Note: Usually 36 inches deep × 30 inches wide is deemed adequate.
- Have all aluminum conductors had grease applied to minimize oxidation?
- Have exterior rated ground clamps been used at all exterior locations?
- Are all panel boxes clean of debris?
- Are the grounding electrode conductor connections to rods undamaged?
- Are panels/breakers rated at or above fault condition?
- Are panels sited properly? Note: Usually panels should be set from ≥48 inches to ≤75 inches above grade.

---

[3] Authority Having Jurisdiction (AHJ): NFPA's definition of AHJs includes officials, agencies, departments, and organizations. Generally, AHJs have two responsibilities: enforcement and approval, that is, AHJs have the official authority and duty to enforce compliance with a standard or code, and to approve the use of systems, strategies, practices, procedures, protocols, plans, methods, machines, facilities, and installations.

[4] OSHA has codified requirements in 29 CFR 1926 Subpart K; they are nearly verbatim replication of the NEC requirements.

- Has Ground Fault Circuit Interrupter (GFCI) protection been provided for all outlets?
- Do overhead poles comply with size and bracing requirements and are they buried deep enough?
- Does wiring meet code or utility company requirements?
- Are all connections tightened to manufacturer's specifications?
- Is the contractor maintaining good housekeeping practices in and around the temporary installation?

SITE WORK

Site work and landscaping are typically the first and last tasks, respectively, on a building site. If the site work is done wisely, the contractor can increase the value and reduce the cost of landscaping after construction. Usually, neither the contractor nor the inspector will have any say in how the buildings are to be situated. But experienced contractors and inspectors may be able to influence the amount of site disturbance that occurs and protect existing trees. They may even affect such matters as future on-site solar and water recovery.

The building permit, drawings, and specifications and, if required, the geotechnical report should be available to show building department inspectors.

GENERAL

- Has a site survey been conducted to identify sensitive areas and features to be protected, such as wetlands, trees, and other vegetation? Notes: (1) If trees are healthy and not too close to the structures, it may be wise to protect them. Mature trees add value to most properties, and appropriately placed trees can reduce a facility's conditioning needs significantly. (2) A tree's root system extends quite a distance from the trunk; even compacting the soil can harm the roots. Therefore, a fairly large area around each protected tree will have to be fenced off, which may work against your desire to save the tree.
- Have engineered fill materials and *in situ* compaction tests been conducted? Note: It is common that a table of compaction test reports will be required if new structures are built on engineered fill or project fill placement exceeds 500 cubic yards. The compaction curves that were utilized for engineered fill placement should be available on site.
- Is there documentation that confirms adequate footing embedment and inspection of pier drilling? Note: A simplified foundation plan should be available.
- Are inspection reports that show that subgrade materials were adequately prepared, all subdrains and retaining wall back-drains were properly installed, basement and crawl space excavations were properly constructed, and so forth, available for a building inspector?

SOILS AND FOUNDATION WORK

Have adequate keying and benching of fill materials been accomplished so that the in-place materials can be constructed upon?

- Are temporary and final cut slopes stable and without any signs of adverse structural conditions?
  - ☐ Is site drainage such that erosion is prevented and that excess surface water or groundwater is collected? Note: Handling storm water runoff can be a major design issue. Contractors and inspectors may be able to recommend solutions that are softer than the usual hard drainage structures. These solutions may include pervious surfaces to allow rainwater infiltration directly into the ground. Use of swales for rainwater instead of curbs and storm water drains may provide both better environmental and less expensive solutions for the site work. Some of these solutions may fall under the Value Engineering clause of the contract, if there is one.
- Have foundations been embedded properly?
- Are final pier depths in compliance with plans? Note: Justification for acceptance of reduced pier lengths is necessary.
- Were subgrade materials, capillary breaks, and vapor barriers properly installed?
- Were subdrains and backwater drains properly installed?
- Were cleanouts and discharge locations properly installed?
- Were the basement excavations as expected based on the soils report? Note: Unexpected changes should be documented.
- Were crawl spaces constructed so as to avoid water from accumulating?
- Were utility trenches properly prepared to receive piping?

BASEMENT/CELLAR[5] WALLS, FOOTINGS, AND SLABS

This inspection is usually conducted after the footings/foundations have been formed, but prior to the placement of concrete.

GENERAL

- Are required materials on site and are they protected?
- Are the footing trenches free of water, debris, and clumps of dirt?
- Are setbacks easily verified?
- Do the grading, drainage, and sizes of the earthwork conform to the plans and specifications?
- Do drains discharge by gravity or mechanical means into an approved drainage system?
- Have all vegetation, topsoil, and foreign debris been removed from the work area?
- Are the footing excavations properly located and are the bearing soil conditions uniformly solid?

---

[5] Using New York City's definitions, a cellar is partly or wholly underground, but having one-half or more of its clear height (measured from finished floor to finished ceiling) below the curb level. Cellars are usually not counted as stories in measuring the height of buildings. On the other hand, a basement is partly underground, but having less than one-half of its clear height (measured from finished floor to finished ceiling) below the curb.

- If subsurface water is uncovered, are special actions necessary to relieve these situations?
- Have all footing beds been prepared with the proper granular fill materials?
- Where footing bearings are on undisturbed soil, will they support at least 2000 psf?
- Is all bearing soil flat (not exceeding a 1:10 slope in the footing trenches) and is it firm and consistent?
- Are the concrete reinforcing bars (rebar) properly installed in walls and footings?
- If applicable, are winter weather or excessive heat procedures in place?
- Have concrete truck clean-out locations been identified?
- With regards to masonry, have the footings for masonry fireplaces and their chimneys been constructed of concrete or solid masonry and are they properly founded and sized? (see Mechanical Rough-in)

CONCRETE SLABS

Inspectors should ensure that preparations for concrete basements, house slabs, and other slabs have been completed prior to delivery of the first batch of concrete. Generally, the following rules apply to footings and slabs: the concrete should be a minimum of 2,500 psi and 3,000 psi if exposed; reinforcement, if required, should be a minimum of two #4 rods that are supported on approved supports; all joints should have an overlap of at least 15 inches, the wet setting of rods is not permitted, and bulk mixing of concrete on site is not permitted.

- Are concrete floors ≥3½ inches?
- Is sand and washed gravel fill in place?
- Is plumbing stubbed through the fill?
- Is termite treatment completed?
- Has 6-mil polyethylene plastic sheeting (Visqueen[6]) been laid? Notes: (1) Leave sufficient vapor retarder (Visqueen) to lap over footings and seal to foundation walls. (2) Overlap joints 6 inches and seal with a sealant (manufacturer's tape). (3) Some designers omit the Visqueen on purpose believing it is not necessary or may do harm, in which case as the inspector, go with the design.
- Have drainpipes been stubbed through the slab?
- Have copper water lines been stubbed up and covered to prevent contact with the concrete?
- Are footing sizes and alignments in accordance with the plans?
- Are footings located below the frost line (where required) or at a depth that is required by local code?
- Are all footing corners square at the bottom of the trenches?
- Is all drain tile, washed stone, filter fabric, sump rock installed, and connected per plans? Note: French drains should be provided around foundations that enclose habitable or useable spaces.
- Are perimeter drainage systems in place?
- Are all forms for slabs the correct thickness?

---

[6] Visqueen is a brand of polyethylene plastic sheeting produced by British Polythene Industries Limited and has become a generic description for any plastic sheeting.

- Has concrete been ordered as per specifications?
- Are garage floors sloped to facilitate the movement of liquids?
- Have all other plan requirements been completed?
- Are the under-slab utilities in place?
- Are all drains properly sized, fitted, and sloped?
- Are public utility connections properly installed as per the plans and specifications?
- Have utility connections, for example, water, sewer, gas, and electrical, been inspected by the utility company inspectors?
- Have tests been properly documented?
- Have all utility relocations been properly documented and recorded?

CONCRETE FOUNDATION WALLS

The foundation supports the entire house and, therefore, a failure can make a structure uninhabitable. Inspectors should look for moisture and water that may require expensive repairs even before the structure has been accepted. Look for uneven settlement that will distort the structure's frame and eventually literally pull it apart. Usually a single localized failure can be corrected by re-leveling of beams or floor joists. Eventually almost all foundations will show some settlement cracks.

Keeping water away from the foundation is the first step in keeping a dry and solid basement, thus the checklist question about downspouts; this can usually be accomplished by ensuring drainage is sloped from the structure. If drainage doesn't solve the problem, sump pumps and waterproofing compounds can be applied to the inside walls to reduce water penetration.

- Are all wall thicknesses as shown on the plans? Note: Usually a minimum thickness of 8 inches is required.
- Have horizontal and vertical reinforcement been properly placed and tied?
- Do steel dowels anchor the foundation walls to footings? Note: A key in the footings provides additional resistance to lateral slippage.
- Have the tops of the foundation walls been prepared to receive, support, and anchor the building walls and floor systems to the superstructure? Notes: (1) Foundation walls that support beams or girders should have footings that are separate from the floor slabs. (2) Foundation walls that support beams or girders should have ≥4 inches solid masonry under all beams or girders.
- Are there positive anchorages installed to resist lateral, uplifting, and overturning forces?
- Have premolded fillers and sealants been placed between floor slabs and walls?
- Do the foundation walls provide for a minimum of ½-inch air spaces on the tops, sides, and ends of wood beams that enter the wall or will the wood be treated?
- If metal floor joists are used, have steel base plates been anchored to the concrete foundation wall?
- Are the walls straight in length and height? Note: Too much pressure may cause walls to bulge inwardly during backfilling operations.
- Do downspouts direct water away from buildings? Note: Rainwater should not terminate into French drains.

CONCRETE MASONRY FOUNDATION WALLS

The previous comments about concrete foundation walls apply here.

- Are anchor bolts for sill plates in place? Are they for light frame construction?
- Are cells in the top course filled with grout?
- Are masonry units laid with a running bond and have proper mortar types been use?

RECOMMENDED GUIDE FOR THE SELECTION OF MORTAR TYPES

BUILDING LOCATION	MORTAR TYPE
Exterior, above grade	
Load bearing	N or S or M
Non–load bearing	N
Parapet wall	N or S
Exterior, at or below grade	
Interior	S or M
Load bearing	N or S
Non–load bearing	N

- Do steel dowels anchor the foundation walls to footings? Note: A key in the footings provides additional resistance to lateral slippage.
- Are there full mortar joints or have the footings been roughened at the base of the wall?
- Have the top of the foundation walls been prepared to receive, support, and anchor the walls and floor systems to the superstructure?
- Is there positive anchorage installed to resist lateral, uplifting, and overturning forces?
- Have premolded fillers and sealants been placed between floor slabs and walls?
- Do the foundation walls provide for a minimum of ½-inch air spaces on the tops, sides, and ends of wood beams that enter the wall or is the wood treated?
- If metal floor joists are used, is there a continuous bond beam on the top of the wall and have steel base plates been anchored to it?
- Are concrete masonry block foundation walls with >4 feet of backfill properly reinforced per the International Residential Code?

WOOD FOUNDATIONS

Common practice for wood foundation walls requires that (1) backfills not be placed until the basement floor and first floor have been constructed or the walls have been braced; (2) backfill material should be gravel or crushed stone for drainage; (3) perimeter footing drainage should be installed; (4) wood foundation walls should not be backfilled until the basement floor and first floor have been constructed or the walls have been braced; and (5) perimeter footing drainage should be installed.

- Have Type 304 or 316 stainless steel fasteners been used where required? Note: Some codes also allow for hot-dipped galvanized steel, silicon bronze, or copper.
- Has all lumber and plywood been pressure-preservative treated and dried?
- Does all lumber bear the label of an accredited agency?
- Has 6-mil polyethylene sheeting been placed between the wall and surrounding ground? Notes: (1) The polyethylene material should be lapped at least 6 inches and bonded with a sealant (manufacturer's tape). (2) A wood strip that extends ≥2 inches above and ≥5 inches below the finish grade should be placed completely around the wood foundation to protect the polyethylene material from ultraviolet light and mechanical damage.
- Are there continuous 1-inch strips between the wood walls and interior concrete slabs?

ELECTRICAL ROUGH-IN[7]

The next three sections are Electrical, Mechanical, and Plumbing Rough-in. If you as an inspector have a say in the matter, you should walk through the project with the subcontractors to ensure coordination between them. Usually the ideal sequence is plumbing waste, mechanical, plumbing supply and then electrical; in any event get agreement between the subs, as they will be working almost one on top of the other. Get the subs to layout their systems before they run them. For the electrical, the layout sequence is usually can lights first, then the other lights, and then the switches. Make sure all work is measured carefully and if in doubt check with the designer.

Electrical, mechanical, and plumbing rough-in inspections are made before insulation, sheet rock, paneling, or other materials cover the work. Most jurisdictions use the National Electrical Code as modified to meet local needs.

GENERAL

- Has the wiring been run to all locations? Note: If the meter base and the service panel are not located back-to-back or next to adjacent stud cavities a four wire system flexible conduit or cable sheath should extend not less than ¼ inch inside the box and beyond any cable clamp.
- Has rigid, nonmetallic conduit been used in corrosive areas?
- Is all equipment and bussing free of paint?
- Are wiring methods (usually cable assemblies) suitable for intended use and environmental conditions? Are boxes suitable for use?
- Do cable installations through or parallel to framing members allow for 1¼-inch clearances or protective steel plates? Are protective steel plates of the appropriate length and width?
- Have the panel boxes had the grounds and neutrals made up and have the service entrances been run?
- Are all junction and outlet boxes accessible?

---

[7] References used: 2014 *NFPA 70*®: *National Electrical Code*; and New York City 2011 Electrical Code (Local Law 39 of 2011).

- Are cables secured to boxes? Note: Where single gang nonmetallic boxes are used and cable is fastened within 8 inches of the box, securing to the box is not required.
- Have flexible conduit or cable sheaths been extended to not less than ¼ inch inside the box and beyond any cable clamps?
- Are boxes that are intended to be flush with combustible and noncombustible finished surfaces been properly positioned?
- Have all splicing devices on all equipment grounding conductors within boxes and bonding connections to metal boxes been checked?
- Have boxes been checked for conductor fill?
- Are equipment grounding conductors correctly sized and suitable for their intended use?
- Have boxes that are used in floors, or to support ceiling fans or similar equipment been checked to ensure correct usage?
- Have recessed luminaires been checked for proper clearances from combustibles and insulation? Note: Recessed light housings should be approved for insulated ceilings. They must be IC[8] rated where installed in contact with insulated ceilings.
- Are smoke and carbon monoxide detectors located properly and in accordance with local standards? Notes: (1) Smoke alarms are generally required when interior alterations, repairs, or additions requiring a building permit are being done. (2) The National Fire Protection Association (NFPA) recommends that one smoke alarm be placed on each floor, in every sleeping area, and in every bedroom. In new construction, the smoke alarms should be AC powered and interconnected. For additional coverage, it is recommended that smoke alarms be in all rooms, halls, storage areas, finished attics, and basements, where temperatures normally remain between 40°F and 100°F. (3) Care should be taken to ensure that no door or other obstruction could keep smoke from reaching the smoke alarms. (4) Local codes may also require smoke alarms to be on every level of a home, including finished attics and basements; inside every bedroom, especially if people sleep with the door partly or completely closed; in halls near every sleeping area, and where halls are >40 feet at each end of these halls; and at the tops of first-to-second floor stairways, and at the bottom of basement stairways. (5) The International Association of Fire Chiefs recommends a carbon monoxide detector on every floor of a home, including the basement. A detector should be located within 10 feet of each bedroom door and there should be one near or over any attached garage.
- Has central heating equipment other than fixed electric space-heating equipment been supplied by individual branch circuits?

---

[8] If a light fixture is intended for direct contact with insulation, it will require an IC rating (IC stands for Insulated Contact). An IC-rated fixture must, by definition, "be approved for zero clearance insulation cover by an OSHA NRTL laboratory," such as Underwriters Laboratory. If the space that a light fixture is to be installed does not contain insulation, a NON-IC–rated fixture should be used (NON-IC stands for NON Insulated Contact). If insulation is present in an application where a NON-IC–rated fixture is used, a minimum 3-inch clearance should exist on all sides of the fixture, and no insulation may be present across the top of the installed fixture. NON-IC–rated fixtures are seldom used in residential applications, but are found in commercial applications; most residential, single-family dwellings will use insulation in the attic space for energy conservation.

- Does the wiring within a manufactured building, including branch circuit wiring, comply with the local codes?
- Do all 480V 1200A electrical rooms have two exits? Note: Two exits are required unless: (1) the working clearance is double what is required, or (2) there is a "continuous and unobstructed way of exit or travel (NEC 110.26)."
- Is all insulation damage free?
- Do feeders to buildings have disconnects outside or immediately inside, where conductors pass through?
- Has enough wiring been installed for future growth?

KITCHENS

- Have wall and countertop receptacles, including islands, peninsulas, and areas behind corner-mounted ranges and sinks been properly spaced?
- Are all required receptacle outlets supplied by small-appliance branch circuits?
- Have a minimum of two 20-A small-appliance branch circuits been used for kitchen receptacles? Notes: (1) Not fewer than two small-appliance branch circuits must supply kitchen countertop receptacle outlets. (2) In dwellings with more than one kitchen, no small-appliance branch circuit can serve more than one kitchen. (3) Receptacles for cord-and-plug connected range hoods should be supplied by individual branch circuits. (4) The two or more small-appliance branch circuits should have only receptacle outlets in the pantry, dining room, and breakfast room, and an electric clock receptacle and electric loads associated with gas-fired appliances. (5) In a kitchen, receptacle outlets cannot be used as lighting outlets.
- Are wall-switched lighting outlets provided and are they wired on a general lighting circuit?
- Have properly sized circuits been provided for specific kitchen appliances, such as dishwashers, disposals, ranges, cooktops, trash compactors, and the like?

DINING ROOMS

- Are receptacle outlets spaced properly?
- Is the wall-switch–controlled lighting outlet on a general lighting circuit?
- Are all required receptacle outlets supplied by small-appliance branch circuits?

BATHROOMS

- Have the receptacle outlets been installed on the wall or partition adjacent to and ≤ 36 inches of each basin? Notes: (1) A receptacle need not be mounted on the wall or partition if it will eventually be installed on the side or face of a basin cabinet. (2) Receptacles are not permitted within or directly over a bathtub space or shower stall. (3) Receptacle outlets cannot be used as lighting outlets in bathrooms.
- Are receptacle outlets supplied by dedicated 20-A branch circuits? Note: If a 20-A circuit supplies a single bathroom, this circuit may supply other outlets (e.g., lighting) within the same bathroom.

- Has GFCI[9] protection, accessibility, and bonding of electrical equipment and grounded metal parts been installed for special equipment, for example, a hydro-massage tub? Note: This branch circuit will typically be a dedicated branch circuit per the rating of the motor.
- Are wall-switch–controlled lighting outlets on a general lighting circuit?
- Are ceiling fans and other hanging fixtures located ≥3 feet horizontally from rims of tubs and showers?
- Have fans with metal exhausts been installed and vented to the outside?

OTHER INTERIOR SPACES

- Are receptacle outlets spaced properly?
- Are bedroom wall-switch–controlled lighting outlets that supply 120 volts (e.g., receptacle, lighting, smoke detectors) Arc Fault Circuit Interrupter (AFCI)[10] protected on all branch circuits?
- Do hallways have at least one wall-switch–controlled (or automatic-, remote-, or centrally controlled) lighting outlet?
- Do hallways with a continuous length of ≥10 feet have at least one receptacle outlet?
- Do stairways have at least one wall-switch–controlled or automatic, remote, or centrally controlled lighting outlet?
- On stairways that include an entry way and where there are six or more risers between levels, are wall switches provided at each floor level and landing level?
- Are clearances between luminaires that are in closets and storage areas in accordance with the relevant code (probably the NEC)?
- Has at least one receptacle outlet been installed for each laundry?
- Are dedicated 20-A circuits supplying all laundry outlet(s)? Is there a laundry receptacle outlet within 6 feet of each intended appliance location?
- Have proper branch-circuit conductors, including equipment grounding conductors, been installed for 240-V dryers?
- Are the lighting outlets for the laundry areas supplied from general lighting circuits?
- Is there at least one wall-switch–controlled lighting outlet and at least one receptacle outlet in each garage?

BASEMENTS AND ATTICS

- Is there at least one receptacle outlet provided in each unfinished basement area in addition to any receptacles that are installed for the laundry?
- Has a receptacle outlet been installed near each installed piece of mechanical equipment? Note: Near is usually defined as being ≤25 feet.
- Have individual branch circuits been installed for central heating equipment?

---

[9] A Ground Fault Circuit Interrupter (GFCI) is an electrical wiring device that disconnects a circuit whenever it detects that the electric current is not balanced between the energized conductor and the return neutral conductor. This imbalance may indicate current leakage through a person who is grounded and accidentally touching the energized part of the circuit.

[10] An Arc Fault Circuit Interrupter (AFCI) is a circuit breaker designed to prevent fires by detecting an unintended electrical arc and disconnecting the power before the arc starts a fire.

Note: The NEC allows for auxiliary equipment and permanently connected air conditioning equipment to be connected to the same branch circuit.

- Have wall-switch–controlled lighting outlets or lighting outlets containing a switch been installed at the entrance to equipment that will require servicing?
- Do accessible attics, attic entrances, and scuttle holes have proper clearances or protection from cable assemblies?

OUTSIDE AREAS

- Are there at least two receptacle outlets that are accessible at ground level, one in the front and one in the back of the house?
- Where there is mechanical equipment, has a receptacle outlet been installed within 25 feet?
- Have cable assemblies (such as Type NM cable) been properly sleeved through concert, brick, and so forth?
- Are wiring methods suitable for their intended use and environmental conditions?

MECHANICAL ROUGH-IN[11]

FORCED-AIR FURNACES

- General:
  - □ Are proper clearances provided for all installed heat-producing equipment?
  - □ Are the platforms upon which the heat-producing equipment is set sufficiently large?
  - □ Have receptacle outlets been installed near all installed mechanical equipments? Notes: (1) "Near" is usually defined as being ≤25 feet. (2) Switch-controlled lighting should be provided for the servicing of equipment. (3) Also on electrical rough-in inspection checklist.
  - □ Are switches or circuit breaker near to disconnect motor-driven appliances?
- Under floors:
  - □ Are the access openings to equipment and passageways in under floor areas large enough to remove the largest piece of equipment? Note: The code may require minimum dimensions.
  - □ Are switch-controlled luminaires located at accesses to the spaces in which equipment is located?
  - □ Are furnaces that are installed in under floor areas suspended ≥6 inches above grade or installed on a slab that extends above an adjoining grade?
  - □ Do clearances that surround equipment comply with applicable codes?

---

[11] References used: The 2006 International Residential Code; the 2006 International Fuel Gas Code (IFGC); the 2006 Uniform Plumbing Code (UPC); the 2005 National Electrical Code (NEC); the 2006 Ventilation and Indoor Air Quality Code (VIAQ); the 2006 Washington State Energy Code (WSEC); and, InspectAPedia®, Free Encyclopedia of Building & Environmental Inspection, Testing, Diagnosis, Repair.

Note: Clearance requirements vary depending on the material upon which the equipment is founded.

□ Are radiant slabs thermally isolated from the soil with a minimum of R-10 insulation that is approved for the use? Note: Where radiant heat systems are being installed, mechanical rough-in inspections must be conducted prior to the slab/foundation pours.

- Attics:
  □ Are the access openings and passageways to equipment in attic areas large enough to remove the largest piece of equipment? Notes: (1) Most codes set minimum dimensions for attic access, for example, 20 inches by 30 inches and located within 6 feet of the equipment. (2) It is common for codes to require a working platform that is the full length of the unit, 30-inches-wide and with a 30-inches-high clear working space. See also the section titled Moisture, Thermal Protection and Insulation, Access hatches and doors that follows.
  □ Is all duct work in attics insulated? Note: Codes usually set a minimum requirement, for example, R-6.

- Garages:
  □ Is all equipment that has a flame, generate a spark or use a glowing ignition source open to the space in which it is installed, and is it elevated such that the source of ignition is above the floor as per code? Note: Usually sources of ignition are required to be ≥18 inches above the floor.
  □ Are ducts that penetrate a wall or ceiling that separates the garage from the dwelling of sufficient gauge and with no openings to the garage? Note: Usually 26 gauge is required.
  □ Have bollards or wheel stops been installed where equipment is subject to impact by an automobile?
  □ Do garage-mounted, vertical air handlers on platforms have return air ducts? Note: Generally drywall boxes are not acceptable.

CONDENSING FURNACES (HIGH EFFICIENCY)

Is condensate drained by gravity to an approved drain or condensate pump? Note: Usually the minimum requirement is for a ¾-inch drainpipe with a ⅛-inch/feet slope.

OIL BURNERS[12]

- Are the oil burners accessible?
- Is the air supply adequate to assure continuous complete combustion? Note: Generally, about one square inch of un-louvered (unobstructed) combustion air intake is required per 1000 btuh of the oil-fired heating boiler, furnace, or water heater.
- Are chimney connectors secure and have the joints been fastened?
- Are barometric draft regulators provided?
- Are inside tanks ≤660 gallons? Note: This may vary, check local code.

---

[12] Primary reference for this section: Wolfeboro Fire-Rescue Department, Oil Burner Inspection Checklist that is based on the New Hampshire State Fire Code/NFPA 31 2001 Edition.

- Are fuel tanks ≥5 feet from burner or source of combustion? Note: Automatic shut-off valves should be installed at tanks and burners.
- Is all piping and tubing supported and protected?
- Are outside fill pipes ≥24 inches from openings that are at the same or lower level?
- Are outside vent pipes ≥24 inches horizontally and vertically from any opening? Note: Vent pipes should be ≥1¼-inch vent pipe.
- Are there remote switches outside entrances to boiler/furnace rooms?
- Are service switches in view of flames? Note: Automatic switches should be installed over burners.
- Are chimney connectors ≥18 inches from combustibles or properly protected? Note: This includes connectors that lead to power vents.
- Are masonry chimneys lined?
- If an oil burner is in a garage in which motor vehicles can be parked, is it ≥18 inches above the slab and is it protected from vehicle contact?
- If a fuel tank is inside a garage, is it protected from vehicle contact?
- With regards to power vent units, are vent intakes and terminations ≥12 inches above finished grade?
- Are vent termination ≥48 inches from any building opening?

DUCTING

- General:
  - Is all ductwork to code and manufacturer's recommendations? Typical requirements include:
    - Are there two story maximum vertical rises on factory made ducts?
    - Are the duct-to-ground clearances ≥4 inches?
    - Are ducts that are in or under concrete encased in ≥2 inches of concrete?
    - Are round ducts with crimped joints lapped a minimum of 1½ inches and are they fastened with three sheet-metal screws or rivets that are equally spaced around the joint?
    - Are joints, seams, and fittings of ducts sealed with tape, mastic, or other approved means?
    - Does ducting, including enclosed stud bays or joist cavities that are used to transport air, that is installed outside of conditioned spaces have all seams and joints sealed?
    - Is flex duct supported per manufacturer's specifications?
    - Are metal ducts supported at least every 10 feet?
- Return air:
  - Is the return air that is taken from a room or space ≥25% of the total volume served? Note: Return air cannot be taken from a bathroom, kitchen, toilet room, mechanical room, closet, furnace room, other dwelling unit, or a garage.
  - Are return air inlets located ≥10 feet from any fuel-burning appliance fire box, or draft hood that is located in the same space?
  - Is the return air duct size ≥2 square inches/kBtu output rating of the furnace or as otherwise specified by the manufacturer?

- Insulation:
  - ❏ Are ducts, boots, and connectors that are in unconditioned spaces, cement slabs, and in the ground and are used for heating or cooling properly insulated?
  - ❏ Is all ducting that is installed in cold walls properly insulated?
  - ❏ Is exhaust ventilation ducting properly insulated?

COMBUSTION AIR

- Is ducting in cold walls insulted as per code?
- Do duct dimensions comply with code?
- Dampers are not permitted in combustion air ducts. Has this requirement been met?
- Are combustion air intakes at least ≥10 feet from return air outlets?
- Are openings sleeved to ≥6 inches above ceiling joists and insulation?
- Ducts cannot be screened when terminating in an attic. Has this requirement been met?
- Where combustion air is obtained from attics, the attics must be sufficiently vented. Has this requirement been met?

If a building is of unusually tight construction, air must be drawn from outside the building. Has this requirement been met? Note: Outside combustion air openings should be screened.

- If a building is of ordinary tightness, combustion air may be drawn from inside the building if the conditioned space is >50 cubic feet per 1,000 Btu/h input for all fuel-burning appliances combined. Has this requirement been met? Note: Buildings that are extremely tight and, thus, do not let much outside air to enter are governed by other rules; check the local code.
- Have all design considerations regarding foundation, under floor, vertical, horizontal, and combinations thereof supply ducts been considered?

VENTS AND CONNECTORS

Gas vent and connector requirements vary greatly depending on the size of the gas vents, horizontal distances from obstructions, roof slopes, and so forth. Refer to the applicable code and manufacturer's listings and performance standards for each specific situation. Some design issues to consider include required roof clearances that depend on the size of the gas vents; horizontal distances from obstructions; clearances from combustibles; wall penetration requirements; vent terminations for power and direct venting; roof slopes; proper sizing when two gas appliances are vented through a common vent connector; vent terminations for power and direct venting; proximity to gas meters and fuel tanks; proximity to property lines and adjacent buildings; proximity to interior corners; vents that pass through attics; and vent and chimney connectors that are installed in the same space. Make sure venting systems are adequately supported.

APPLIANCES

It is hard to imagine a house being built in the United States without appliances. Here checklists are provided for clothes dryers, gas ranges, fireplaces, and air conditioning. The section on Plumbing Rough-in addresses dish washers and water heaters.

- Clothes dryers: Clothes dryers are a common cause of house fires; many of these fires result in loss of life and property. Most of these fires are the result of improper lint cleanup and maintenance. Fortunately, they are very easy to prevent if the dryers are installed properly and then maintained.
  - ☐ Have dryers been installed as per manufacturer's instructions?
  - ☐ Are all clothes dryer exhaust ducts of metal with smooth interior surfaces and with joints running in the direction of air flow?
  - ☐ Are connectors visible, that is, not concealed in construction?
  - ☐ Screws should not be used to attach connectors to ducts. Has this requirement been met?
  - ☐ Have duct connectors been sized properly? Note: Usually duct connectors should be ≥4 inches or the appliance outlet size.
  - ☐ Are clothes dryer ducts ≤25 feet? Note: Codes have formulas to account for bends in ducts, for example, deduct 2.5 feet for each 45° bend and 5 feet for each 90° bend. See also manufacturer's installation instructions.
  - ☐ Are clothes dryer exhaust ducts independent of other ducted systems?
  - ☐ Are there backdraft dampers with no screens at the exterior termination points of clothes dryer ducts?
  - ☐ Are exhaust duct terminations that are on the outside of buildings ≥3 feet in any direction from openings into any building?
  - ☐ Do the clothes dryer transition ducts comply with Underwriters Laboratories UL 2158A "Clothes Dryer Transition Duct?" Note: This duct is an approved standard for flexible high-temperature exhaust ducts that are rated to 430° F. They can be used on both electric and gas dryers.
  - ☐ Do gas connectors have shutoff valves installed immediately ahead of the connectors?
  - ☐ Gas connectors should not be concealed within, or extend through walls, floors, partitions, ceilings, or appliance housings? Are these requirements met?
- Gas ranges and range hood ducts:
  - ☐ Are the vertical clearances to combustibles at least 30 inches or as per manufacturer's listing?
  - ☐ Are all gas connectors ≤6 feet long.
  - ☐ Are all gas shutoff valves installed immediately ahead of the connector?
  - ☐ Do range hood ducts terminate outside of the house?
  - ☐ Are range hood ducts air tight and are they equipped with a backdraft damper?
  - ☐ Are range hood ducts made of galvanized steel, stainless steel, or copper, and do the ducts have smooth interiors?
- Fireplaces: Bad things can happen with a malfunctioning fireplace. Chimney fires spread quickly to roof structures and cause major damage. A new fireplace should be carefully inspected and tested by an inspector knowledgeable of fireplaces before being turned over to a new homeowner.
  - ☐ Are factory-built fireplaces and wood burning stoves listed and labeled? Note: The Environmental Protection Agency (EPA) certifies factory-built fireplaces and wood-burning stoves for their burning efficiency and particulate.

- Are hearth extensions easy to distinguish from the surrounding floor, and are they in accordance with the fireplace listing?
- Are all penetrations sealed with listed materials?
- Have spark arresters been installed?
- Have shutoff valves that are located in fireplace fireboxes been installed as per the manufacturer's instructions?
- Have all decorative shrouds that are used at chimney terminations been properly listed and labeled for use with their intended chimney systems?
- Are flues centered over fireboxes to avoid uneven drafting?
- Are there allowances at damper ends for expansion and contraction?
- Are fire stop spacers placed at each floor or ceiling?

FIRE STOP SPACER

- Are fireplace foundations for masonry fireplaces and chimneys large enough to spread the load evenly and so as not to exceed the soils carrying capacity?
- Do all un-insulated spaces between chimneys and combustible walls and ceilings meet code? Note: Usually a space of ≥ 18 inches is required.
- Air-conditioning:
  - Are cooling coils downstream (return side) from the heat exchanger?
  - Have adequate workspaces been established around each air conditioner? Note: Many codes establish minimum work areas.
  - Do all condensate lines drain to approved places of disposal? Note: Draining under a house and into a crawl space is not allowed.
  - Are condensate lines a minimum of ¾-inch diameter and do they slope without sagging?
  - Have the refrigerant lines been pressure tested? (Compressors should not be started prior to a successful pressure test.)

EXHAUST VENTING

- Are ventilation fans in kitchens, bathrooms, water closet rooms, laundry rooms, and indoor swimming pools and spas source specific?
- Are bathroom fans rated at ≥ 50 cfm?
- Are kitchen fans rated at ≥ 100 cfm?
- Are exhaust ducts that terminate outside of the building equipped with backdraft dampers and insulated to a minimum of R-4 in unconditioned spaces such as attics and crawl spaces? Note: Codes may vary regarding the insulation rating.

WHOLE HOUSE VENTILATION SYSTEMS

- Intermittent whole house ventilation using exhaust fans:
  - ☐ Do the whole house fans that are located ≤4 feet from the interior grille have a sone[13] rating of ≤1.5?
  - ☐ Are remotely mounted fans acoustically isolated from structural members and solid ductwork?
  - ☐ Are there easily accessible 24-hour timers tied to the exhaust fans? Note: Timers should be set to operate as per code.
  - ☐ Are controls properly labeled?
  - ☐ Are there air inlets of ≥4 square inches in each habitable room?
  - ☐ Are doors undercut a ≥½ inch where separated from the exhaust source? Note: Exhaust-only ventilation systems do not require outdoor air inlets if the home has a ducted, forced-air heating system that communicates with all habitable rooms and the interior doors are undercut ≥½ inch above the finished floor covering.
- Whole house ventilation integrated with a forced-air system:
  - ☐ Are screened outdoor air inlets connected to return air plenums that have motorized dampers?
  - ☐ Are outdoor air inlet duct connections to the return air stream located ≤4 feet upstream of the forced-air blower?
  - ☐ Are there easily accessible 24-hour timers tied to furnace blowers and motorized dampers? Note: Timers should be set to operate as per code.
  - ☐ Are controls properly labeled?
- Intermittent whole house ventilation using supply fans:
  - ☐ Are supply fans inline?
  - ☐ Is outdoor air filtered before it is delivered to habitable rooms?
  - ☐ Are outdoor supply-side inlets located downstream of blowers?
  - ☐ Are outdoor return-side outlets ≥4 feet upstream of blowers?
  - ☐ Are there easily accessible 24-hour timers tied to the inline supply fan? Note: Timers should be set to operate as per code.
  - ☐ Are labels reading "Whole House Ventilation (See operating instructions)" affixed to controls?
- Whole house ventilation using a heat recovery ventilation system:
  - ☐ Are ducts ≥6 inches in diameter?
  - ☐ Are balancing dampers installed on both the inlet and exhaust side?
  - ☐ Are supply ducts in conditioned spaces upstream of heat exchangers insulated to meet code?
  - ☐ Are there easily accessible 24-hour timers tied to the inline supply fan? Note: Timers should be set to operate as per code.
  - ☐ Are labels reading "Whole House Ventilation (See operating instruction)" affixed to controls?

---

[13] A *sone* is a measurement of sound of exhaust fans in terms of the comfortable hearing level for an average listener. The lower the sone value, the more comfortable the listening environment will be. One sone is equivalent to the sound of a quiet refrigerator in a quiet kitchen. Typically, the sone level is measured at maximum cubic feet per minute (speed); however, some newer products are also being tested at normal CFM settings to provide consumers with typical sound level information.

- Outdoor air inlets:
  - ☐ Are inlets screened?
  - ☐ Are inlets located so as not to draw air from any of the following:
    - An appliance vent outlet that is ≤10 feet, unless the vent outlet is ≥3 feet above the outdoor air inlet?
    - A place where it will draw in objectionable odors, fumes, or flammable vapors?
    - A hazardous or unsanitary location?
    - A room or space that has any fuel-burning appliance in it?
    - ≤10 feet of a vent opening for a plumbing drainage system unless the vent opening is ≥ß 3 feet above the air inlet?
    - An attic, crawl space, or garage?

PLUMBING ROUGH-IN[14]

In this section specifics that were extracted primarily from the Uniform Plumbing Code are provided. This was done to give you an idea of what inspectors may check, but you need to check your local code to ensure correctness; local authorities may alter the Uniform Plumbing Code (UPC). For example, one question asks, "Trap arms that are ≥3 inches in diameter and that change direction by ≥135° require cleanouts when they are installed at the lowest level of a gravity drain." A local code may say "change direction by ≥180°, instead of 135°."

UNDERGROUND PLUMBING

- Is piping that passes under or through walls protected from breakage?
- Are the voids that surround pipes that pass through concrete floors that are on ground sealed?
- Is piping that passes through or under cinders or other corrosive materials protected from external corrosion?
- Have provisions been made for hot water piping and piping for solar systems to expand, contract, and settle? Note: Solar piping should not be directly embedded in concrete, unless designed and listed for such use.
- Have sleeves been provided to protect piping through concrete and masonry walls and concrete floors? Note: Sleeves are not required where openings are drilled or bored.
- Are vent and branch vent pipes free from drops or sags, and is each such vent level or graded and connected so as to drip back by gravity to the drainage pipe it serves?
- Are vents downstream of traps?
- Has Type L copper tubing or better been used for underground water lines? Note: Type M copper tubing may be used for water piping when piping is above ground in, or on, a building.

---

[14] References used: The 2006 Uniform Plumbing Code (UPC); the 2006 International Residential Code (IRC); 2005 National Electrical Code (NEC); Idaho Administrative Code, Division of Safety, IDAPA 07.02.06, "Rules Concerning Uniform Plumbing Code," Division of Building Safety; the 2006 International Fire Code (IFC); and, Residential Water Heater Checklist, City of Palos Verdes 'CA' Community Development Department.

- Have all ferrous metals been wrapped?
- Have water lines been tested to working pressure or 50 psi for 15 minutes? Note: Plastic pipe is not allowed to be air tested.
- Has building sewer piping been laid on a firm bed? Note: Where piping is laid on made or filled-in ground materials, have the materials and method of support been approved as required by the authority having jurisdiction?

SEWAGE EJECTORS

- Wherever possible, do plumbing fixtures drain to a public sewer or private sewage disposal system? Note: Only fixtures that are on floor levels below the crown[15] level of the sewer should discharge through an ejector.
- Is the drainage piping that serves fixtures that have flood-level rims located below the elevation of the next upstream manhole cover of a public or private sewer protected with an approved backwater valve? Note: Fixtures that are above such elevation shall not discharge through the backwater valve.
- Are the discharge lines from the ejectors provided with accessible backwater or swing check valves and gate or ball valves?
- Do the sewage ejectors' or pumps' discharge capacity comply with code?
- Do pumps have audio and visual alarms and are they accessible?

DRAINS

- Have drains, wastes, and vents (DWV) been water tested with a 10-feet head for 15 minutes or air tested at 5 psi for 15 minutes? Note: Plastic pipe is not allowed to be air tested.
- Are drains properly sized?
- Where two fixtures are set back-to-back, or side-by-side, and are served by a single vertical drainage pipe, does each fixture's waste go into an approved double-fixture fitting with inlet openings at the same level?
- Do horizontal drainage lines that connect with other horizontal drainage lines enter properly?
- Do vertical drainage lines that connect with horizontal drainage lines enter properly?
- Are tub waste openings into crawl spaces that are at or below the first floor closed off with metal collars or metal screens that are fastened to the adjoining structure, with openings no greater than ½ inch in the least dimension?
- Are cleanout fittings and cleanout plugs/caps of an approved type? Are cleanouts gas and watertight?
- Have 18-gauge nail plates been installed where plastic or copper plumbing is within 1 inch of the face of framing?
- Are plastic lines supported ≤4 feet? Is there a support at each horizontal branch connection?
- Are vertical plastic lines supported at their base and at each floor? Have midstory guides been provided?
- Are vertical cast iron hubless pipes supported at their base and at each floor (not to exceed 15 feet)?
- Are waste pipes protected from freezing?

---

[15] The very top of the inside of a pipe is its crown.

- Are treaded ABS (acrylonitrile-butadiene-styrene) fittings and joints accessible?
- Are fittings for drains and vents directional type?

TRAPS

- Does each plumbing fixture have a trap? Note: Traps in fixtures that are not used regularly or are seldom used may dry out thereby allowing odors to enter the building. You may suggest pouring water down these drains periodically.
- Are bathtub traps accessible? Note: Consider installing drum traps; cylindrical traps that are closed on the bottom and have a cover plate for access.
- Is each trap protected by a vent?
- Do the developed lengths of the installed trap arms[16] comply with code?

UNIFORM PLUMBING CODE

PIPE SIZE DIAMETER	LENGTH OF ARM TRAP
1¼	2′ 6″
1½	3′ 6″
2	5′
3	6′
≥4	10′

- Do trap arms that change direction have cleanouts? Notes: (1) Trap arms that are ≤3 inches in diameter and that change direction by more than 90° require cleanouts. (2) Trap arms that are ≥3 inches in diameter and that change direction by more than 135° require cleanouts when they are installed at the lowest level of a gravity drain.
- Are the vertical distances between fixture outlets and their traps as short as practicable? Note: Vertical distances between fixtures and traps should not exceed 24 inches in length except for clothes washers that may have a maximum 30-inch standpipe.

CLEANOUTS

- Does each horizontal drainage pipe have a cleanout at its upper terminal? Notes: (1) Cleanouts are not required at horizontal runs ≤5 feet except at sinks. (2) Cleanouts may be omitted on any horizontal drainage pipe that is installed on a slope of ≤72° from the vertical angle. (3) Cleanouts are not required above the lowest level of the gravity drain. (4) An approved two-way cleanout fitting that is installed inside a building wall near the connection between the building drain and building sewer or installed outside of a building at the lower end of a building drain and extended to grade may be substituted for an upper terminal cleanout.
- Is each run of piping ≥100 feet in total developed length[17] provided with a clean out for each 100 feet or portion thereof?

---

[16] A trap arm is that portion of a fixture drain that is between the trap weir and the vent pipe fitting.

[17] The UPC uses the term "developed length" quite often. Developed length means the physical length of the pipe including all offsets or bends from the point of origin to the point of termination. The point of origin for a vent is the point where it attaches to the drain line it serves.

- Are there cleanouts located at each aggregate horizontal change of direction exceeding 135°?
- Are all cleanouts installed so that they open to allow cleaning in the direction of flow or at right angles to the flow and, except in the case of a wye branch and end-of-line cleanouts, installed vertically above the flow line of the pipe?
- Are under floor cleanouts ≤20 feet from access doors with an unobstructed 30-inches-wide x 18-inches-high pathway to them?
- Are all cleanouts accessible? Have they been extended above floors or outdoors where access is limited?

ISLAND-FIXTURE VENTING

An island-fixture vent, aka "Chicago Loop" is used where conventional vertical vent stack or air admittance is not feasible or allowed.

- Are the top elbows at least as high as the peak possible drain water level in the sinks that they serve?

ISLAND-FIXTURE VENT

Source: Diagram from Wikipedia.

- Is island-fixture venting used only for sinks and lavatories? Note: Kitchen sinks with a dishwasher waste connection, a food-waste grinder, or both, in combination with the kitchen sink waste are permitted.
- Are the trap arms at least two times the pipe diameters? (For example, a 1¼-inch-diameter pipe requires at least a 2½-inch trap arm.)
- Are there cleanouts on both the vents and the drains?
- Are there connections to vertical drainpipes or to the top half of horizontal drain pipes?

VENTS

- Are vent pipes made of cast iron, galvanized steel, galvanized wrought iron, copper, brass, Schedule 40 ABS DWV, Schedule 40 PVC DWV, stainless steel 304 or 316L, or other approved materials having a smooth and uniform bore? Note: Galvanized wrought iron, galvanized steel, and stainless steel 304 pipes and fittings cannot be installed underground and must be kept at least 6 inches above ground.
- Do copper tubes that are used for drainage and vent piping have a weight of not less than that of copper drainage tube type DWV?
- Are approved fittings used wherever there is a change in direction of vent piping?
- Are vent pipes sized properly?
- Do all vent pipes or stacks extend through flashing and do they terminate vertically ≥6 inches above the roof (10 inches in high–snow load areas) or ≤1 feet from any vertical surface?
- Do all vents terminate ≥10 feet from or ≥3 feet above any operable window, door, opening, air intake, or vent shaft, or ≥25 feet from any air intake or vent shaft?
- In snow areas with design temperatures below 0°F, are vent terminals ≥2 inches in diameter, but in no event smaller than the required vent pipe?
- If there are any combination waste and vent systems, have they been specifically approved by the authority having jurisdiction?

AIR ADMITTANCE VALVES (AAVs)[18]

- Are AAVs ≥4 inches above the crown of the trap?
- Are AAVs ≥6 inches above insulation in attics?
- Are AAVs accessible and open to air flow?

AIR ADMITTANCE VALVE (AAV)

---

[18] Air admittance valves, also called AAVs, Durgo valves, and Studor Vents are negative pressure–activated, one-way mechanical vents that are used to eliminate the need for conventional pipe venting and roof penetrations. A discharge of wastewater causes the vent to open, thereby releasing the vacuum and allowing air to enter the vent pipe for proper drainage. Some state and local building departments prohibit their use, but the International Residential and International Plumbing Codes allow them in lieu of vents that penetrate roofs.

WET VENTS

Wet venting is commonly used in conjunction with toilets and sinks; the drain for the sink is also the vent for the toilet; when used, inspectors should familiarize themselves with the numerous rules that pertain to wet vents.

- Are all wet-vented fixtures on the same story?
- Are wet vents ≤6 feet in developed length?
- Have wet-vented sections been properly sized? Note: Wet vents are sized based on the fixture unit discharge into the wet vent. The minimum size of a wet vent is 2 inches for 4 drainage fixture units (dfu) or less, and 3 inches for more than 4 dfu.
- Do wet vents serve a maximum of four fixtures?

WATER SERVICES

- In buildings with both potable and nonpotable water, are systems clearly marked: green background with white lettering for potable water and yellow background with black lettering, with the words "Caution: Nonpotable water, do not drink," for nonpotable water?
- Where there is reclaimed water, is the system marked with purple (Pantone color #512) background and imprinted in nominal ½-inch high, black uppercase letters, with the words "Caution: Reclaimed water, do not drink?"
- Are backflow preventers installed as required? Note: Look especially at fire-sprinkler systems, irrigation systems, and other sometimes overlooked cross-connections.
- Is the water service buried deep enough to protect it from freezing?
- Have the meter and building supply pipes been properly sized? Note: Normally, no building supply pipe should be less than ¾ inch in diameter.
- Is the static water pressure in the water supply piping ≤80 psi? Note: If the water pressure exceeds 80 psi, you will have to install regulators that control the pressure to all water outlets unless otherwise approved by the authority having jurisdiction.
- Is it possible to turn the water off to each building being supplied by the system? Are shutoff valves accessible?
- In multidwelling units, is there one or more shutoff valves per dwelling unit? Are these valves easily accessible in the dwelling unit that they control?
- Are full-way[19] valves installed to control all outlets on the discharge side of each water meter and on each unmetered water supply?
- Are dielectric fittings or other approved fittings used between different metals?
- If a metal pipe water service has been replaced with a plastic pipe water service, does the building's grounding system comply with applicable plumbing and electrical codes?
- Have proper materials been used? Note: The International Plumbing Code (IPC) accepts all of the common water piping materials, such as copper

---

[19] A full-way valve, also known as a gate valve, is a flow control device that consists of a wedge-shaped gate that can be raised to allow full, unobstructed flow or can be lowered to restrict the flow passage. A full-way valve is not intended for close fluid flow control or for very tight shutoff.

tubing, CPVC, galvanized steel, cross-linked polyethylene, and PEX-AL-PEX; however, some jurisdictions may exclude some materials.

- Have water lines been tested to working pressure or 50 psi for 15 minutes? Note: Plastic pipe is not allowed to be air tested.
- Have ford fittings or other fittings that are deemed equal been installed on plastic water pipes?
- Are water pipes in different trenches than the building's sewer and drainage pipes and are they laid on approved materials? Notes: Exceptions may be made if: (1) the bottom of the water pipe, at all points, is at least 12 inches above the top of the sewer or drain line, or (2) the water pipe is placed on a solid shelf that was excavated at one side of the common trench with a minimum clear horizontal distance of at least 12 inches from the sewer or drain line. (3) Water pipes that cross sewer or drainage pipes that are constructed of clay or materials that are not approved for use within a building must be laid a minimum of 12 inches above the sewer or drainpipe.
- Is the water piping that was installed within the building[20] and in or under a concrete floor or slab resting on the ground installed properly? Note: Refer to the code on this issue (currently UPC 609.3). Generally, ferrous piping should have a protective coating of an approved type, machine applied and conforming to recognized standards. Field wrapping to provide equivalent protection is restricted to short sections and fittings necessarily stripped for threading. Galvanized coating is not deemed adequate protection for piping or fittings. Approved non–ferrous piping need not be wrapped.
- Has underground copper tubing been installed without joints wherever possible? Where joints were necessary, are they brazed and are the fittings of wrought copper?
- Do valves, including pressure-reducing valves, that are installed in the ground have access boxes?
- Where metallic water services have been replaced with metallic water pipe, was the grounding system replaced properly? Similarly, if the replacement was with plastic pipe, was a proper grounding system installed?

WATERLINES

- Have water hammer arrestors or air chambers been installed? Have the devices been installed as per manufacturer's specifications for location and installation? Note: Codes vary on this issue and should be checked.
- Are exterior water lines properly insulated?
- Have water lines been tested to working pressure or 50 psi for 15 minutes? Note: Plastic water piping cannot be tested with air.

GAS PIPING

- Are all exterior metallic gas piping and fittings coated with corrosion-resistant material?
- Where more than one gas meter is installed, has each gas meter been identified with a brass tag?

---

[20] Within the building indicates within the fixed limits of the building foundation.

- Have test gauges located at the gas stub out location near the gas meter?
- Has new piping been pressure tested? Note: Tests generally require 10 psi on a 15-psi gauge for 15 minutes.
- Have gas meters been verified to meet demand?
- Do all fittings and supports conform to code?
- Have gas meter clearances been verified? Note: Check to ensure that meters are located in accordance with utility and local requirements: distance from electric meters, vents, and operable window and so forth. Windows, building vents, cable television (CATV), telephone, electric panels, and other sources of ignition must be ≥18 inches horizontally or ≥10 feet vertically from the gas regulator spring case vent, unless the regulator is installed in an enclosure with a regulator vent extension.
- Is all underground metal and plastic piping installed with ≥ 18 inches of cover? Note: Exceptions may be permitted.

WATER HEATERS

- Tank water heaters:
  - Where required, have two seismic straps been installed on each water heater; one located within the top 1/3 of the water heater unit and one at the bottom 1/3? Note: The bottom strap must be ≥4 inches from water heater controls.
  - Are all vents and the water heaters clear from combustible materials? Note: Generally this clearance should be ≥6 inches when the water heater vent material is double-walled. Also check the manufacturer's specifications.
  - Are all single-walled vents and single-walled to double-walled vent joints secured with a minimum of three sheet metal screws, rivets, or similar connections?
  - Does each water heater have a pressure/temperature (P/T) valve? Note: The valve must be drained to the exterior and terminate toward the ground, maintaining between ≥6 inches and ≤24 inches from the ground and pointing down.
  - Is the diameter of P/T valve opening maintained to the termination of the drain?
  - When water heaters are located such that a leak could cause damage to underlying wood framing, are the water heaters set in a pan that is constructed of watertight corrosion-resistant material?
  - Are all water heaters that are located in garages elevated so that pilot lights and controls are ≥18 inches above the garage floor?
  - When water heaters are located in bedrooms, bathrooms, or closets, are they provided with a listed self-closing, gasketed door and do they obtain all their combustion air from outdoors?
  - When water heaters are located in attics, are they accessible through a opening and passageway that is at least as large as the largest component of the water heater? Note: Some codes require minimum opening and working platform dimensions and minimum lighting.
  - Where required, has seismic strapping been installed?
  - Where water heaters are installed in the normal path of vehicles, has protection in the form of a wheel stop, bollard, or by elevating been installed?

- Tankless water heaters:
  - ☐ Are all manufacturer's instructions available for inspectors?
  - ☐ Are all tankless water heaters listed by an approved testing agency (e.g., UL, UPC) and are they installed in accordance with manufacturer's specifications? Note: check especially gas supply lines and venting material requirements.
  - ☐ Are both gas and water shut-offs adjacent to the water heater? Notes: (1) All new gas piping must be pressure tested. (2) Most codes set a maximum length for gas flex connectors.
  - ☐ Have expansion tanks been installed wherever the water system is "closed loop" (when the water cannot move back into the water supply system)? Note: A system is considered a closed loop when a water regulator and or backflow prevention device is installed.
  - ☐ Do vents provide combustion air per manufacturer's requirements?
  - ☐ Are all tankless water heaters independently vented with a category III (Stainless steel UL 1738 certified) venting system or per manufacturer's specifications? Note: Tankless water heater vents have restrictive clearance to combustibles requirements.
  - ☐ Are condensate drains installed on horizontal runs if required?
  - ☐ When tankless water heaters are installed in attics and furred spaces, have pans with ¾-inch drain pipes been installed under the heaters?
  - ☐ When tankless water heaters are installed on interior walls, are the walls protected from water leakage with sheet metal at water heater location to the floor?
  - ☐ Has fire blocking at ceiling and floor levels been verified?

LAUNDRIES

- Are traps for clothes washer standpipe receptors installed above floors?
- Are traps for clothes washer standpipes roughed in between ≥6 inches and ≤18 inches above the floor? Note: Standpipe receptors for any clothes washer shall extend between ≥18 inches and ≤30 inches above their traps.
- Have water hammer arrestors or air chambers been installed? Have the devices been installed as per manufacturer's specifications for location and installation?
- Have floor drains been placed in laundry rooms in commercial buildings and common laundry facilities in multifamily dwelling buildings?
- Has all dryer duct work been installed and properly vented to exterior? Note: Generally, dryer duct length in a residence cannot exceed 25 feet with a deduction of 5 feet made for each 90° elbow and 2½-feet deduction for each 45° elbow.

KITCHENS

- Do all dishwasher drains have air gaps[21]?
- Have water hammer arrestors or air chambers been installed? Have the devices been installed as per manufacturer's specifications for location and installation?

---

[21] Air gap: A fixture that provides back-flow prevention for an installed dishwasher. Air gaps are usually located above the countertop as a small cylindrical fixture mounted parallel with the faucet. Below the countertop, the drainpipe of the dishwasher feeds the "top" of the air gap, and the "bottom" of the air gap is plumbed into the sink drain below the basket or into a garbage disposal unit. Air gaps prevent drain water from the sink from backing up into the dishwasher.

BATHROOMS

- Are faucets and shower head fittings properly supported?
- Have ADA requirements been met where required?
- Are water closets set to code? Note: Usual requirements are ≥15 inches to center from a sidewall with a total clear width of ≥30 inches and ≥21 inches at the front.
- Do toilet rooms with two or more water closets or a combination of one water closet and one urinal have a floor drain? Note: Dwelling units are exempt from this requirement.
- Are over-rim tub faucets set with ≥1-inch air gap[22] to the tub rim?
- Have water hammer arrestors or air chambers been installed? Have the devices been installed as per manufacturer's specifications for location and installation?
- Are dams to shower stalls ≥2 inches to ≤9 inches?
- Are doors to shower stalls ≥22 inches wide?
- Do shower stall floors drain properly?
- Are shower pans properly constructed? Note: Most codes are quite specific as to shower compartment construction.

EXTERIOR

- Are there vacuum breakers on all hose faucets? Note: There should not be any valves downstream of the vacuum breakers.
- Has backflow protection been provided on all irrigation systems?

FRAMING[23]

There are several approaches to framing a house: (1) Balloon framing that utilizes studs that run the full height of the frame from the sill plate to the roof plate. Joists are nailed to the studs and supported by sills or by ribbons let to the studs. (2) Platform framing that uses studs that are one story high. They are placed one on top of the other when there are more than one story. (3) Post and beam construction utilizes a framework of vertical posts and horizontal beams that carry both the floor and roof loads; the floor and roof loads are transmitted to the posts (columns) that transmit the loads to the foundation. In each case there are code issues of which you should be aware.

A framing rough-in inspection is usually done following the electrical, mechanical, and plumbing rough-in inspections. Sheetrock should not be stacked against walls as this may impede the inspection.

GENERAL

See also High-rise construction, Wood and Plastics, Carpentry and millwork.
- Has the lumber been graded or certified by an approved agency?

---

[22] A few words about air gaps: With regards to a drainage system, air gaps are unobstructed vertical distances through air between a wastewater pipe outlet and the flood-level rim of the receptor into which it is discharging. An air gap prevents wastewater from backing up into the originating wastewater pipe. With regards to water-distribution systems, air gaps are the unobstructed vertical distance through air between the lowest opening from a water supply discharge to the flood-level rim of a plumbing fixture.

[23] References are: The 2006 International Residential Code (IRC) and the 2006 Washington State Energy Code (WSEC).

- No significant moisture should remain in the wood framing; is this the case?
- Have all subtrade rough-ins, for example, electrical, plumbing, and mechanical, been completed? Has any rough-in work caused damage?
- Is the nailing as per code and per plan?
- Have shear walls been installed as per schedule? Note: Check especially for specified connections at walls, plates, joists, and so forth.
- Have anchor bolting been installed per the shear wall schedule?
- Are properly sized nuts tightened on each bolt?
- Are all doors and windows installed per approved plans? Has tempered glazing been installed at all the required areas, such as tubs, showers, stairs, walkways, doors, and adjacent areas? Notes: (1) Where the opening of an operable window is located ≥72 inches above the outside finished grade or surface below, the lowest part of the clear opening of the window should be ≥24 inches above the finished floor of the room in which the window is located. (2) Usually for a residence, at least one exterior door is required to be ≥36 inches wide by ≥80 inches high. (3) Emergency exiting cannot be through a garage.
- Are anchor bolts properly fastened to the sill plate and do they line up properly?
- Has decay protection been installed? Note: Decay protection is generally an exception, not a rule. Inspectors should ensure that sound construction practices have been followed. For example, there should be adequate drainage for the site and buildings, flashing should not be skimped, there should be proper clearances between ground and wood, crawl spaces and attics should be well ventilated, and vapor barriers should be installed as required. It is wise to learn and inspect against local experience and to recommend control measures that match the local environment. Where subterranean termites are a problem, soil treatment chemicals may be required.
- Are all joists, beams, and girders bearing as required and are they fastened properly?
- Have improper notches been cut or holes been bored into beams? Have roof trusses been cut improperly to make room for HVAC equipment?
- Are required metal straps installed and are they the appropriate type? Note: Often flat straps are bent around the beams, which is not a proper installation.
- Is all stair framing properly aligned? Are all requirements regarding such matters as vertical rise, stair width, stair nosing, open risers, riser rises, radius of curvature at the leading edge of treads, landing area, and headroom been met?
- Has other framing been improperly modified in other ways? Note: This is often done to install wiring and plumbing.
- Does attic access allow for HVAC equipment servicing? Note: Local codes vary at the required opening dimensions.
- Are joists supported laterally?
- Have fire blocking and fire-stopping been installed as required? Have penetrations at the top and bottom plates, fire blocks, soffits, ceiling lines, and so forth, been sealed? Notes: (1) Fire blocking is generally required at soffits, chases, dropped ceilings, and stairs where the underside is unfinished, and so forth. (2) Un-faced fiberglass insulation that is used as fire blocking must fill the entire cross-section of the wall cavity to, a minimum height of, generally, 16 inches vertically. (3) Insulation must be packed tightly around obstructions

such as piping, conduits, and so forth. (4) Mineral wool or glass fiber used for fire blocking must be securely retained.

- If there is a crawl space, has it been properly ventilated? Is there access to the crawl space?

WALLS

- Does bearing at floor joists meet code? Note: Bearing requirements differ for wood, steel, and masonry and concrete.
- Are top plates overlapped with other partitions at the corners and intersections?
- Are framed openings per code? Note: See also the Plumbing Rough-in and the Mechanical Rough-in checklists for additional requirements for access within the crawl spaces and attics.
- Are windows and glass door assemblies properly anchored as per manufacturer's recommendations?

ROOFS

- Are roofs complete and have exterior moisture barriers been installed?
- Have roofs been properly framed and are they secured?
- Are attics properly ventilated?
- Are ridges, hips, and valleys designed as per code?
- Do notches on the ends, tops, and bottom of rafters comply with the code?
- Are rafter ties completed?
- Does the roofing material conform to the original design?

TRUSSES

- Are the truss specifications on site? Have the truss specifications been sealed by a professional engineer? Note: Make sure premanufactured trusses are certified (stamped) by the manufacturer.
- Does the truss configuration conform to the design drawings?
- Do trusses have bearings as per specifications?
- Are required hangers installed per specifications?
- Are connection plate sizes, gauges, and locations per specifications?
- Has all truss bracing been completed per specifications?
- Are ganged trusses nailed off as per manufacturer's specifications?

ENERGY REQUIREMENTS

- Have the building thermal envelopes[24] been durably sealed per local code requirements? Notes: (1) Often an Energy Code Checklist must be

---

[24] 2012 International Energy Conservation Code (IECC) defines the building thermal envelope as follows: "The building thermal envelope is the barrier that separates the conditioned space from the outside or unconditioned spaces. The building envelope consists of two parts—an air barrier and a thermal barrier that must be both continuous and contiguous (touching each other). In a typical residence, the building envelope consists of the roof, walls, windows, doors, and foundation. Examples of unconditioned spaces include attics, vented crawl spaces, garages, and basements with ceiling insulation and no HVAC supply registers.

completed and submitted at the final Inspection. (2) Typical minimum energy requirements are R-18 exterior walls. R-38 ceiling/attic spaces, R-21 floors over crawl spaces, and R-10 basement walls (conditioned); however, some jurisdictions require larger R values.

- Do windows have a maximum U-factor of 0.40 and a Solar Heat Gain Coefficient (SHGC) of 0.40[25]?
- Does each window have a label? Note: Labels should not be removed before building department inspections. Labels should certify the U-Factor and SHGC and that the air leakage requirements are met.
- Do doors whose surface area is more than one-half glass comply with the requirements for windows?
- Are manufactured windows and exterior doors certified as meeting an air infiltration rate not exceeding 0.3 cfm/ft^2 of window or door area?
- Are windows and exterior doors weather-stripped and have all joints and penetrations been caulked and sealed?

MOISTURE, THERMAL PROTECTION, AND INSULATION[26]

Vapor retarders do what their name implies: retard the flow of moisture vapor. They are made of low-permeable materials and are installed to prevent moisture from entering and reaching the dew point within the building. In cold and temperate climate, they are usually placed as close to the warm side of insulated construction as possible. In warm and humid climate, the vapor retarder is usually placed closer to the outer face of the construction.

GENERAL

- Is the roof complete and are exterior moisture barriers properly installed?
- Are roofs, especially flat roofs and low-sloping roofs, capable of carrying all possible snow loads?
- Has flashing been installed at critical locations: roof edges, changes in slope, around parapet walls, chimneys, vents, skylights, and the like? Were the flashing materials chosen properly?
- Does all insulation comply, R-rating wise, with local code requirements?
- Does all insulation with facings, vapor barriers, or breathable papers have the required flame spread rating? Note: The International Residential Code requires insulation materials, including facings, such as vapor retarders or vapor-permeable membranes that are installed within floor-ceiling assemblies, roof-ceiling assemblies, wall assemblies, crawl spaces, and attics to have a flame-spread index ≤25 with an accompanying smoke-development index ≤450 when tested in accordance with ASTM E 84. These limits do not apply when the facing is installed in substantial contact with the unexposed surfaces

---

[25] Solar Heat Gain Coefficient (SHGC) is a major energy-performance characteristic of windows: the ability to control solar heat gain through the glazing. Solar heat gain through windows is a significant factor in determining the cooling load of many commercial buildings.

[26] References are: The 2006 International Residential Code (IRC), the 2006 Washington State Energy Code (WSEC), and Ching, Francis D. K., "Building Construction Illustrated, fourth edition.

of ceilings, floors, or walls. Un-faced insulation is acceptable when concealed in areas previously mentioned.

- Is insulation securely installed? Look especially at fireplaces, crawl spaces, and other out-of-sight places.
- Are recessed incandescent luminaires thermally protected and listed as thermally protected? Notes: (1) Insulation is not required when the luminaires are installed in poured concrete. (2) Insulation is not required when the luminaires have design, construction, and thermal characteristics that are equivalent to insulated luminaires.
- Is insulation continuous at foundation, floor, wall, and roof junctures so that there is an unbroken envelope against heat transfer? Note: Exterior joints around windows, door frames, openings between walls and foundations, openings at utility services through walls, floors, and roofs should be sealed, caulked, gasketed, or weather-stripped to limit air leakage.
- Are insulation clearances in accordance with manufacturer's specifications? Is loose insulation held in place with barriers?

ACCESS HATCHES AND DOORS

- Are access doors from conditioned spaces to unconditioned spaces weather-stripped and insulated to match surrounding surfaces?
- Are attic access hatches installed so as to match and maintain the R-values at the accesses?
- Is all weather-stripping material durable under extended use, noncorrosive, and replaceable?
- Are access hatches large enough to get HVAC and other equipment through? Notes: (1) Most codes specify minimum hatch dimensions. (2) Access to equipment should allow access without damaging insulation.
- Where required, have self-latching, self-closing doors with UL-rated hardware been installed?
- Are exterior doors and doors serving as access to enclosed unheated areas weather-stripped?

ATTIC INSULATION

- Are eave and cornice vents installed so that insulation does not block the free flow of air? Note: A minimum of 1-inch space should be between the insulation and roof sheathing and the location of the vent.
- Are attic vents constructed so as to keep rain and snow from entering attics?
- Are R-value markers placed in attics? Note: Markers that show the thickness and maximum settling thickness of loose insulation should be spaced throughout attics, with at least one marker for every 300 square feet.

WALL, FLOOR, AND CEILING INSULATION

- Is all faced insulation stapled over the face of framing members? Note: Insulation may be un-faced if a Visqueen vapor barrier is installed over the whole wall or a Polyvinyl Acetate (PVA) primer[27] is used to seal drywall.

---

[27] Polyvinyl Acetate (PVA) primer is a resinous high polymer that is colorless, water soluble and thermoplastic that is used as latex. It provides a tight bond to dry wall.

- Are vapor barriers installed on the warm side of walls, floors, and ceilings? Notes: (1) Moisture vapor, a gas, always migrates from high- to low-pressure areas. Therefore, it tends to flow from high-humidity to low-humidity areas. In winters this is usually from inside to outside and in summer from outside to inside, especially if the air is air-conditioned. (2) Vapor barriers are not required in roof/ceiling assemblies where the ventilation space above the insulation is ≥12 inches.
- Does insulation have at least 1 inch of vented air space above it? Note: Vapor retarders are not required where (1) all of the insulation is installed between the roof membrane and the roof deck and (2) where the vented space above the insulation averages ≥12 inches.
- Are bottom plates and corners at the insides of exterior walls been caulked?
- If faced batts are used as insulation, is the insulation face stapled?
- Is the floor insulation in contact with the floor that it is insulating?
- Is insulation properly supported? Note: This will also be checked during the final inspection.
- Are air vents to crawl spaces open and not blocked by insulation? Notes: (1) Baffles may be installed at 30° from the horizontal to direct airflow to the lower surface of the insulation. (2) As an alternative to insulating floors of crawl spaces, insulation of the crawl space walls is allowed when the crawl space is not vented to the outside.

SLAB INSULATION (ON-GRADE AND BELOW GRADE)

- Are 6-mil Visqueen sheets or other approved vapor retarders with joints lapped ≥6 inches placed between concrete slabs and base courses or prepared subgrades? Note: See Vapor Retarders, discussed later.
- Where slabs are on-grade or less than 12 inches below grade, does the insulation extend downward from the top of the slab on the outside or inside of the foundation wall? Note: Other code restrictions generally apply.
- Is exposed above-grade insulation protected from physical and ultraviolet damage?
- Are conditioned basement walls insulated from the top of the basement wall down 10 feet or to the basement floor, whichever is less? Note: Walls in unconditioned spaces must meet the same requirement unless the floor overhead is properly insulated.
- Are radiant slabs thermally isolated from the soil with a minimum of R-10 insulation that is approved for the use? Notes: (1) Where radiant heat systems are being installed mechanical rough-in inspections must be conducted prior to the slab/foundation pours. (2) The same question was asked under Mechanical Rough-in.

VAPOR RETARDERS

- Are vapor retarders installed and are they made of materials that comply with code?
- In framed structures where the walls, floors, and roof/ceilings make up elements of building thermal envelopes, are the vapor retarders installed on the warm-in-winter side of the insulation? Notes: Exceptions are (1) where

moisture or freezing will not damage materials; (2) where the framed cavity or space is ventilated to allow moisture to escape; (3) areas specifically identified as exceptions; and (4) where all of the insulation is installed between the roof membrane and the roof deck.

- Has 6-mil Visqueen been installed in all crawl spaces? The Visqueen should be overlapped 12 inches and should run wall to wall.

ONE-COAT STUCCO

One-Coat Stucco can be used for exterior and interior surfaces. The exterior applications are usually for residential and light commercial use. Because it has a thinner base coat that is applied in a single layer, application costs may be reduced over the traditional three-layer system. One-Coat Stucco is fiber reinforced; thus, shrinkage is minimal and there tends to be less cracking than in the three-layer system.

- Does the base have a minimum thickness of ⅜ inches?
- Does the expanded polystyrene foam that is used behind the stucco base course ≥1 inch?
- Have control joints been installed per the manufacturer's requirements? Note: Control joints relieve stresses that are caused by thermal changes.
- Is moist curing being accomplished per the manufacturer's requirements?
- Is the metal lath that is used for the One-Coat Stucco base galvanized?
- Synthetic stucco/Exterior insulation finishing system (EIFS):[28] **EIFS** was introduced to the United States in 1969 by Dryvit Systems. **EIFS** was primarily used on commercial buildings in the United States until the early 1980s, when it was introduced into the residential building market. EIFS was initially designed as a *barrier system*, meaning that its success depended upon no moisture ever getting into the building envelope. This created a problem in that mold formed when moisture did get in. Drainable EIFS is available but seldom used. Therefore, you must inspect to ensure the system is essentially waterproof.
- Has the EIFS been installed by an approved contractor?
- Are all flashings properly installed?
- Are all exterior openings properly sealed?

FINAL INSPECTIONS[29]

For the most part final inspections will cover the same inspection items as do rough in inspections. Since walls will be closed-in, only things that are visible will be checked, unless the inspector suspects cheating. It is usual for the contractor, owner, or his or her representative and sometimes the architect/engineer to accompany the building inspector during a final inspection. Some contractors may have a few people available

---

[28] EIFS: Exterior Insulation and Finish System.

[29] Primary references are the 2006 Washington State Energy Code (WSEC), 2006 International Building Code (IBC), and the 2006 International Residential Code (IRC). Throughout this section on Light Construction I have drawn heavily on information provided by MyBuildingPermit.com, a service of eCityGov.net; the following cities in the State of Washington are listed: Mill Creek, Mukilteo, Renton, Sammamish, SeaTac, Snohomish County, Snoqualmie, and Woodinville. Additional references are listed at specific sections, for example, Electrical Final Inspection.

to make on-the-spot corrections and thus avoid a reinspection. Sometimes, one city/ county inspector will conduct the final inspection under the assumption that all important subcontractor issues were covered during rough-in inspections. Other times electrical, plumbing, and mechanical trades will be inspected separately, thus the duplication on checklists.

EXTERIORS

- Are house numbers plainly visible from the street or road fronting the property and are they the proper size?
- Are permits and approved plans on-site and accessible to the inspector?
- Is all permit information correct, for example, address, permit number, scope of work, and so forth?
- When separate sewer, septic, and other permits were required, were they signed off prior to the final inspection?
- Have deficiencies that were noted during the rough-in inspections been corrected?
- Have all exterior windows, penetrations, and openings been properly caulked and weather-stripped?
- Are chimney terminations ≥2 feet higher than any portion of a building that is within ≤10 feet and are the terminations ≥3 feet above the highest point where the chimney passes through the roof?
- Are spark arrestors installed on all chimneys?
- Where there is wood siding, is there ≥6 inches between the soil and wood?
- Is surface drainage diverted to a storm sewer or other approved point of collection so as not to create a hazard? Notes: (1) Water should be drained away from all foundation walls at a minimum of 5% within the first 10 feet and slope toward the nearest public right-of-way. (2) Some communities allow dry wells and the like in lieu of draining to a public right-of-way.
- Unless open on at least two sides, do carports meet the fire separation standards required of garages?
- Are all required hand and guardrails installed?
- Are rain-gutters with downspouts and splash blocks or equivalent over all exterior exits?
- Are slopes greater than 2 to 1 retained properly?
- Are building exteriors properly painted and caulked?
- Has a permanent power inspection been completed and required meters installed prior to the final inspection?
- Are exit doors that are side-hinged, ≥36 inches wide, and ≥6-feet 8-inches high?
- Are there 36-inch landings at all exterior doors? Note: Doors may open at landings that are not more than 7¾ inches lower than floor level, provided the door, other than an exterior storm or screen door does not swing over the landing.
- Have "antisiphon"-type hose been installed where required?
- Are there screens on all operable windows?
- Where required, have all broken sidewalks and curbs been replaced?
- With regards to decks and walkways, have the following criteria been met?

- Are all setbacks, dimensions, and materials as per the approved permits?
- Have ledger strips[30] for the decks been fastened as per approved plans?
- Are decks positively attached and supported both vertically and horizontally? Note: Attachments to houses cannot be toenailed or subject to withdrawal.
- Are deck connections with cantilevered framing members designed and constructed to resist uplift?
- Are cantilevers blocked at bearing when ≥12 feet?
- Are decks made of treated lumber? Notes: (1) Cuts, notches, and holes must be treated with preservative. (2) Untreated joists may be allowed if an approved weatherproof decking material is installed. (3) Soffits may be allowed when ventilated.
- Are there safety grates over all window wells located in a patio or adjacent to a walkway?
- Are fasteners and hardwares of hot-dipped galvanized steel, stainless steel, silicon bronze, or copper?
- Are guardrails installed where required and do they comply with approved plans? Note: Where deck is ≥30 inches above a floor or grade, a 36-inch guard is required.
- Are footings to proper depths? Note: This is an important issue in frost areas.

INTERIORS

- Are all electrical circuits properly identified in panel(s)?
- Has GFCI protection been provided in kitchens, basements, garages, and exterior outlets in conformance with the applicable National Electrical Code.
- Are furnaces properly tagged?
- Are all appliances, HVAC equipment, and other motorized equipment accessible?
- Are smoke and carbon monoxide detectors properly installed? Note: Carbon monoxide detectors should be on each level that has a gas-fired appliance.
- Have water pressure regulators been installed where required?
- Have all finished hardware and floor coverings been properly installed?
- Have AFCI-protected bedroom circuits been installed?
- Does attic access, draft stops, and ventilation conform with the applicable code?
- Where required, has attic insulation been installed and have installers' certificates been posted just inside attic accesses?
- Have all joints and fasteners within garage firewalls been taped and covered by approved means?
- Are water heaters properly installed? See Plumbing rough-in, water heaters, discussed earlier.
- Have the following requirements that pertain to stairways been met? Note: There are numerous stairway types with varying requirements. You will have to check the applicable codes for your exact needs. What follows should be taken

---

[30] Ledger strip: As used here, a horizontal board used for vertical support. A ledger strip would be attached to the house to support the deck.

as guidance as to what building inspectors check. The information provided on stairs is per the International Residential Code; local codes may vary.

- ☐ Headroom: The normal headroom requirement is 6 feet 8 inches, measured vertically from the nose of treads, landings, and platforms.
- ☐ Illumination: A light switch should be at each floor level of six or more risers.
- ☐ Stair nosings: ≥¾ inch to ≤1¼ inch; a nosing is not required where the tread depth is ≥11 inches.
- ☐ Openings: There are numerous requirements regarding the size of openings allowed between open risers, guards, open sides of stairs, and so forth. Check the code for your unique situation.
- ☐ Leading-edge radius of curvature: The leading-edge curvature should not exceed 9/16 inch.
- ☐ Risers: ≤7¾ inches.
- ☐ Treads: ≥10 inches. Note: The maximum deviance between rise and run should not exceed ⅜ inch.
- ☐ Landings: There should be a floor or landing at the top and bottom of each stairway. The floor landing at exit doors should not be >1½ inches lower than the top of the threshold. Numerous other requirements and also exceptions apply to landings at interior and exterior doors. Check the code for your unique situation.

- • Handrails: (1) These are required on a continuous run of four or more risers. (2) Usually, guardrails are required on stairways that are open on one or both sides. (3) The height of handrails should be ≥34 inches to ≤38 inches above the nose of tread to top of handrail. (4) As with openings, the International Residential Code (IRC) has many requirements regarding handrails. This is especially true regarding residences. Check the code for your unique situation.
  - ☐ Has safety glazing been installed in the following locations?
    - – Swinging doors
    - – Fixed and sliding panels of sliding door assemblies and panels in sliding and bifold closet door assemblies
    - – Storm doors
    - – Railings, regardless of area or height above walking surfaces
    - – Unframed swinging doors
    - – Enclosures in doors and enclosures for indoor and outdoor swimming pools, hot tubs, whirlpools, sauna, steam rooms, bathtubs, and showers
    - – Adjacent to stairways, as required.
- • Have the follow requirements that pertain to windows and safety glazing been met (not a complete list)? Notes: (1) Safety glazing is required adjacent to stairways, landings, and ramps ≤36 inches horizontally of a walking surface when the exposed surface of the glass is ≤60 inches above the plane of the adjacent walking surface. (2) Safety glazing is also required adjacent to stairways located ≤60 inches horizontally of the bottom tread of a stairway in any direction when the exposed surface of the glass is <60 inches above the nose of the tread.
  - ☐ Are bedroom windows ≤44 inches from the floor to the bottom of the window opening? Are windows ≥5.7 square feet and do they have a clear opening of ≥20 inches width and ≥24 inches height?

- ☐ Are emergency escape-and-rescue openings operational from the inside without the use of keys, tools, or special knowledge?
- ☐ Has all required safety glazing been marked with type and thickness? Note: Marks should be acid-etched, sandblasted, ceramic-fired, embossed, or made by other permanent means.
- ☐ Has safety glazing been installed in fixed or operable panels that are adjacent to a door where the nearest vertical edge is within a 24-inch arc of either vertical edge of the door in a closed position and where the bottom exposed edge of the glazing is ≤60 inches above the walking surface? Note: An exception occurs where there is an intervening wall or partition between the door and glazing or where the door accesses a closet that is ≤3 feet in depth.
- ☐ Has safety glazing been installed in fixed or operable panels where the exposed area of the individual panel is 9 square feet and the bottom edge is ≤18 inches above the floor and the top edge of the panel is ≥36 inches above the floor? Note: An exception is made where a protective 1½-inch-wide bar is installed on the accessible side of the glazing ≥34 inches to ≤38 inches above the floor and is capable of withstanding a load of 50 lbs per linear foot.
- • Has adequate access been provided to all crawl spaces? Notes: (1) Access openings through floors should be a minimum of 18 x 24 inches, unless local code dictates otherwise. (2) Openings through a perimeter wall should be a minimum of 16 x 24 inches, unless local code dictates otherwise. Through wall accesses should not be below doors to residences.
- • Are crawl space vents free of obstructions, for example, insulation?
- • Have vapor barriers been installed on the ground? Note: Vapor barriers should be black, 6-mil Visqueen (plastic). The plastic should cover wall-to-wall with seams overlapping 12 inches. Note: Some designers omit the Visqueen, in which case so should you.
- • Have minimum insulation requirements been met and is the insulation secure? Note: Typical minimum energy requirements are R-18 exterior walls, R-38 ceiling/attic spaces, R-21 floors over crawl spaces, and R-10 basement walls (conditioned). However, some jurisdictions require larger R values.
- • Has wood been protected from decay? Notes: Wood preservatives are required for the following, per IRC Section R319: (1) Wood joists or the bottom of a wood structural floors when closer than 18 inches or wood girders when closer than 12 inches to exposed ground in crawl spaces or unexcavated area located within the periphery of the building foundation. (2) All wood framing members that rest on concrete or masonry exterior foundation walls and are less than 8 inches from exposed ground. (3) Sills and sleepers that are on concrete or masonry slabs that are in direct contact with the ground, unless they are separated from the slabs by an impervious moisture barrier. Also, the ends of wood girders that enter exterior masonry or concrete walls and that have clearances of less than 0.5 inch on tops, sides, and ends. (4) Wood siding, sheathing, and wall framing on the exterior of buildings with clearances of less than 6 inches from the ground. (5) Wood structural members that support moisture-permeable floors or roofs that are exposed to the weather, such as concrete or masonry slabs, unless they are separated from the floors or roofs

by an impervious moisture barrier. (6) Wood furring strips or other wood framing members that are attached directly to the interior of exterior masonry walls or concrete walls below grade except where an approved vapor retarder is applied between the wall and the furring strips or framing members.

- Are wood columns of approved wood of natural decay resistance or approved pressure-preservative–treated wood?
- Where lumber and plywood are required to be pressure-preservative treated, does it bear the quality mark of an approved inspection agency?
- In areas that are subject to damage from termites, besides the wood having been pressure-preservative–treated wood, has the soil been chemically treated? Note: Both the applicator's qualifications and materials that were used should be verified.
- Are interior posts at basements or cellars 1 inch above floors and 6 inches above exposed earth and are they separated by an approved impervious moisture barrier or are they pressure treated?
- If in a flood hazard area, have structures been designed and constructed to meet this contingency?
- Has all debris been removed from crawl spaces?

FIRE INSPECTIONS

The following is a general list of requirements for business/residential fire inspections:

- Are addresses displayed on buildings and are they visible from the street?
- Are gas meters accessible and protected by a driveway?
- Are fire extinguishers installed and accessible?
- Are all circuit breakers in electrical panels labeled?
- Are electrical outlets not overloaded with power strips?
- Are the required number of smoke detectors installed and are they properly located?
- Are there ≥18 inches of clearance between furnaces and hot water heaters from any combustible materials?
- Is exiting from buildings adequate?
- Are all flammable liquids properly stored?
- With regards to smoke alarms and automatic sprinkler systems, have the following criteria been met? Note: Codes vary greatly from community to community. Here are some common requirements:
  - Have smoke alarms been installed in all new buildings and wherever alterations, repairs, or additions that required a building permit were constructed?
  - Have automatic residential fire sprinkler systems been designed and installed in accordance with both building and fire codes? Note: Systems may have to be installed with a fire department connection (FDC) and other associated devices when required by the fire code official.
  - Are smoke alarms at every floor level, top of stairs, in each bedroom, and in hallways that serve bedrooms?
  - Are all smoke alarms listed and installed in accordance with both the applicable building and fire codes?

    ☐   Where required, has the automatic sprinkler system been approved by the fire department prior to the final building department inspection?

ELECTRICAL[31]

- Are all panels labeled and clear spaces in front?
- Have AFCI/GFCI breakers been properly located and installed?
- Have all GFCI outlets and outlets fed by GFCIs tested?
- Have rod or Ufer[32] grounding/bonding been installed?
- Have emergency lights been installed and do they provide enough light for emergency egress?
- Are exit signs in right locations per plan and have they been tested?
- Have all GFI breakers been tested? Note: Usually, 1000 A, 277/480 V or more breakers require third-party testers.
- Are all disconnects in place with correct fuse sizes and clearances?
- Have all unused openings been closed?
- Are cover plates installed properly? Note: Box positions should be flush with walls and ceilings. If a box is not flush with the finished surface, as a result of improper installation, or if paneling, drywall, tile, or a mirror is added, box extenders, sometimes referred to as a "goof rings," should be installed.
- Are all fire-rated penetrations sealed?
- Is all installed equipment approved and listed?
- Are bubble covers in wet locations?
- Are there receptacles within 25 feet of installed equipment?
- Are there light switches and accessible receptacles by attic accesses?

MECHANICAL

- Are labels and tags affixed as required?
- On whole-house ventilation that uses a heat recovery ventilation system, are there readily accessible, 24-hour timers, set to operate 8 hours/day and tied to inline supply fans?
- Are outdoor air inlets screened?
- Is make-up air provided where range exhaust fans exceed 400 cfm?
- Are fireplaces certified and have they been tested?
- Are washing machines and dryers installed according to code and manufacturer's specifications?
- Have crawl spaces been cleared of debris and is all insulation secure so as not to block air vents?
- Are appliances that have been installed outside listed or protected from outdoor environments?

---

[31] References used: Building and Safety Department, City of Milpitas, CA, Building and Safety Department, Commercial Rough & Final Electrical Inspection Checklist, 2014 *NFPA 70*®: *National Electrical Code*; and New York City 2011 Electrical Code (Local Law 39 of 2011).

[32] The Ufer Ground is an electrical earth grounding that uses a concrete-encased electrode to improve grounding in dry areas. The technique is frequently used in construction of concrete foundations.

- Are sources of ignition on gas appliances (water heaters, furnaces, and electronic air cleaners) within garages ≥18 inches above the floor unless listed as flammable vapor ignition resistant?
- Are exposed ducts within garages a minimum of 26-gauge sheet metal with no openings into the garage?
- Have bollards or wheel stops been installed where equipment is subject to mechanical damage?
- Are all ducts in the attic, garage, crawl space, and other unconditioned spaces, insulated with minimum R-8? Note: This R value varies from place to place.
- Have drip legs been installed at each appliance or where condensation could collect?
- Have unions or flex connectors been installed between shut-off valves and appliances? Notes: (1) Unions or flex connectors should not be concealed within or extend through a wall, floor, partition, or appliance housing. (2) One flex connector up to 6 feet long is allowed on each appliance. (3) Shut-off valves are required for each appliance. They should be located upstream of unions and easily accessible.
- Have sediment traps been installed downstream of appliance shutoff valves? Notes: Shutoffs should be as close to the inlets of the appliances as practical, except for illuminating appliances, ranges, clothes dryers, and outdoor grills.
- Are pipes properly supported?

PIPE DIAMETER	DISTANCE BETWEEN SUPPORTS
½ inch	6 feet
¾ inch–1 inch	8 feet
1¼ inch or larger	10 feet
1¼ inch or larger (vertical)	Every floor level

- Are all supply, return, and exhaust ducts, clothes chutes, chimneys, gas vents, ventilating ducts, dumbwaiters, and elevator shafts free of piping?
- Where there are townhouses, does piping that is installed downstream of the point of delivery extend only to the unit served by the piping? Note: Piping cannot extend through a townhouse.
- Is all vent piping for relief vents and breather vents vented directly and independently to the outdoors? Notes: (1) Vent piping for breather vents can use manifolds in accordance with manufacturer's installation instructions. (2) Vents must be designed to prevent the entry of insects, water, and foreign objects.
- Are gravity appliance venting systems equal in area to the vent collars on the appliances that they serve? Note: Performance standards may reduce the vent sizes.
- Are appliance vents that are connected to a common venting system within the same story at the highest level consistent with headroom and clearances from combustibles? Note: A vent system's area cannot be less than the area of the largest vent plus 50% of the smaller flue collar added.
- Are offsets in gravity appliance vents installed with as many offsets as required that do not exceed 45° from vertical? Notes: (1) No more than one 60° from vertical is allowed. (2) Horizontal runs cannot exceed 75% of the vertical height of the venting system.

- Are vent connectors that serve Category 1[33] appliances *not* connected to any portion of a mechanical draft system that operates under positive pressure?
- Are all vent clearances and terminations in accordance with applicable codes? Note: Allowable distances vary with roof slopes, distances from walls and other vertical objects, distances from entrances, distances from combustibles, and so forth.
- Are all wall penetrations in accordance with applicable codes?
- Where vents extend through insulated assemblies, are insulation shields of the proper gauge and distance from insulation and combustibles?

FURNACES

See Mechanical Rough-in.

PLUMBING

- Are all plumbing installations per schematics and approved plans and specification? Note: It is common that all appliances and fixtures will be inspected during a final inspection.
- Are mains, branches, drains, and vents properly sized?
- Have appropriate fittings been installed?
- Do vents meet all clearance requirements?
- Do hose bibbs have nonremovable vacuum breakers of a self-draining type?

HOSE BIBB VACUUM BREAKER

---

[33] "Category 1 appliance" describes an appliance that uses gas and vents at around an 83% efficiency rate. Due to their efficiency rate, the gas vents at a cooler temperature and has a higher risk of condensation in the vent and connectors. If not properly vented, Category 1 appliances run the risk of leaving acidic condensate. This condensation can be very dangerous and, therefore, their vents are very well regulated. Category 1 appliances tend to be large household appliances such as washing machines, water heaters, and dishwashers.

- Are water heaters properly installed? Notes: (1) If a gas water heater has been installed without having any plumbing pipe modifications, it will be inspected as part of the final mechanical inspection. If an electric hot water heater has been installed or a gas water heater including pipe modifications, it will be inspected as part of the final plumbing inspection. (2) Temperature and pressure relief valves are required on all water heaters. (3) Pipes for the drain should be hard and full sized. (4) Drain pans, seismic straps, thermal expansion devices, and pressure relief valves at all water heaters must be properly installed.
- Have cleanouts been installed as per code? Notes: (1) Check the number of cleanouts, their locations, bends, and slope. (2) Check also the building drain cleanouts/property cleanouts to ensure they are per approved details.
- Are under-floor cleanouts located ≤20 feet from crawl space access doors or trap doors?

2

HIGH-RISE BUILDING CONSTRUCTION

I recall attending a seminar on finance. The speaker gave an example of buying $100 worth of stock, to which a person in the audience yelled out, "Who buys only $100 worth of stock?" The speaker responded, "If $100 is too small an amount just add zeros, the principles remain the same." In a sense the same can be said about moving from light construction to high-rise construction: the buildings are larger and usually more complex, but, as with light construction, high-rise building inspectors must be knowledgeable of construction, building codes, zoning regulations, Occupational Safety and Health Administration (OSHA) regulations, Americans with Disabilities Act (ADA), and so forth. They need to be organized, have good communication skills, both orally and in writing, and be able to work with people. They need to be firm when violations, defects, and poor practices are observed.

However, there are differences between a two-family house and, say, an 800,000-square feet building in Manhattan, New York City that may cost in excess of $400 million. Whereas in light construction, the contractor or superintendent will likely take personal charge and manage the entire job, often also swinging a hammer, the 800,000-feet building will have a staff of highly qualified construction people along with architects, engineers, contracting experts, estimators, schedulers, clerical people, safety personnel, and inspectors who are experts in their field (I've probably forgotten some classification). Nowadays, construction management (CM) firms have overall responsibility for managing these projects, with owners also having a full-time representative in the construction office. Some large corporations have their own CM teams that manage such high-rise and plant projects. So yes, the principles of inspection may be the same, but high-rise construction is different.

It's not possible to provide all the necessary checklists for all high-rise buildings. But I attempt to get you started. Furthermore, even the best plans and specification will have details of construction that are not specifically covered in the plans and specifications. Such situations appear as the work progresses, and it will be through your inspection(s) that proper practices are followed and unsatisfactory materials and faulty methods avoided.

Thus, even with a high-powered CM team on the job, it is important to have inspectors on the job; inspection is an important element of construction and contract administration. Good engineering and properly prepared plans and specifications are essential for a quality end product, and so is ensuring that the specifications and drawings are adhered to.

For openers, here are some areas of responsibility that you will have to take on:

- Maintain complete and accurate records and prepare well-written reports. This requires the completion of numerous documents. These documents must be complete and accurate such that they will stand up in arbitration or a trial. Common among the documents that inspectors must prepare are:
  - □ Inspector's daily logs.
  - □ Project log book (field diaries).
  - □ Photographs and videos for record purposes. All photos and video should be labeled and dated for record purposes.
  - □ Construction progress payments.
- Files including, but not limited, to the following:
  - □ Safety and accident reports.
  - □ Punch (deficiency) lists.
  - □ Contractor technical submittals. Inspectors should verify and document that all required submittals have been approved prior to being installed.
  - □ Contract modifications.
  - □ Payments to contractor. Note: Keep track of actual construction progress against approved schedules.
  - □ Contractor's required notices (posting and maintenance).
  - □ Special reports, for example, suspension of work, and differing site conditions
  - □ Substantial completion and completion reports.
  - □ Final acceptance memoranda and reports.
- Notify the proper authorities in the event unanticipated archeological materials are encountered during construction.
- Maintain liaison with architects, engineers, contracting experts, estimators, schedulers, clerical people, safety personnel, other inspectors, and so forth.
- Ensure that safety standards are being met (unless there is a separate safety structure on the job). Under all circumstances, report unsafe conditions.
- Seek value engineering (VE) opportunities.
- Be alert to issues relating to the ADA.
- Verify that materials and equipment incorporated into the project meet contract requirements.
- Ensure that all construction activities comply with contract requirements and that work is performed in accordance with good construction practices. This requires that inspectors become familiar with all aspects of the project: contract plans and specifications; relevant contract documents; and, knowledge of how the work will progress. It is also essential that the inspector develops a good working relationship with the contractor's personnel and learns about specific instructions that are given to those doing the work.

This last responsibility is at the heart of what you must do, that is, conduct technical inspections. The technical inspections that are conducted and their findings usually drive the other administrative activities that are listed above. Effective inspection and good end products are accomplished by familiarity with contract plans and specifications and relevant contract documents; knowledge of the project work phases; and a good working relationship with the contractor's personnel.

The checklists that follow are intended to assist you in preparing and conducting on-the-job inspections. They assume that you are steadily on the job as construction progresses. Regardless of who pays your wages, it is your responsibility to ensure that the work is done correctly and that the materials that are being installed comply with the specifications. The checklists do not address all possible conditions, include all possible inspection items, or attempt to duplicate complete building codes. Hopefully, they do provide you with an understanding of what to expect as to the breadth and detail of the inspections that you must conduct. And, as with the section on light construction, your inspections will in all likelihood be followed up by building department inspectors.[1]

The following checklists are presented in the approximate sequence as in the Construction Specifications Institute (CSI) document for construction standards. For reference purposes, the CSI division number is in parentheses next to the section title.[2]

SITE CONSTRUCTION

EARTHWORK, EMBANKMENTS, TRENCHING, AND GRADING

Excavating contracts generally fall into the following categories: utility systems trenching, foundation preparation, and waterway construction. OSHA regulations control all trenching operations. Qualified operators should conduct excavation and backfill operations using approved methods and equipment.

- Have controlling elevations been check?
- Have underground utilities been identified, located, staked, and marked? Note: The utility inspection should be completed, identified, and plainly staked/marked on the ground; most states have a toll-free telephone number that must be called before excavation to reduce the likelihood of utility line contact.
- Have all utilities that were not on the project drawings properly documented?
- Have all damaged underground utilities been documented and reported? Note: Usually, it is the contractor's responsibility to repair all damaged utilities.
- Are all open excavations that are over 5-feet in depth properly shored or sloped? Notes: Excavations have several inherent hazards that are not common to other earthwork activities.
- Are open excavations barricaded or posted whenever work is suspended regardless of the time period?
- Is compaction that is within 18-inches of any structure being conducted using hand-operated equipment? Note: Normally heavy rollers and other

---

[1] While on the job you may wish to carry certain basic tools and inspection aids such as a 25-feet tape, 6-feet folding rule, engineer/architect scale, lock level, magnet, high-low thermometer, standard volt-ammeter, and camera; ultimately, the particular project will guide you to what to carry.

[2] In developing these checklists, I have drawn heavily on *Construction Inspection Handbook* (360 FW 4), August 2004, Division of Engineering, U. S. Fish and Wildlife Service, Department of the Interior.

heavy equipment-type compactors are not to be allowed within 18 inches of structures.

- When subsurface conditions differ from the plans, have footings been redesigned by a professional engineer?
- Are foundation materials suitable (free of organic or frozen material, debris, or standing water) for the planned construction? Note: If rock or unsuitable material is present, it must be brought to the attention of the project manager.
- Are all excavated areas filled with an approved compacted backfill or concrete? Note: Once excavation is complete, conduit placement and similar operations can begin. Backfill is placed in layers of designated thickness and then compacted. Once compacted, density testing of the compacted material follows to ensure the complete consolidation of the backfilled material. Once the material is brought up to the designated grade, the surface covering is installed.
- Have all backfill materials that were used under concrete slabs been carefully inspected? Notes: (1) The material must be free of organic and frozen material and debris. (2) The material should be of a type that is readily compactable (certain cases would require that it be free draining). (3) The material, before compaction, must be placed in accordance with contract requirements: It should be brought to specified moisture; and a testing laboratory or a qualified inspector should verify the density and lift thickness. (4) Additional compaction and retesting should be requested as needed.
- With regards to embankments, have all deleterious materials and organic matter been removed from the site? Note: Embankments are usually constructed to either impound water or keep water from entering specific areas. In the context of building construction, the embankments are usually not very large. Nevertheless, considerable damage may occur should any embankment that holds backwater fail. Embankments may also be constructed to gain Leadership in Energy and Environmental Design (LEED) points by protecting or restoring habitat and for storm water quantity control.
- Has the site been scarified so that the soil substrate will be interlaced with the fill materials?
- If required, has dewatering been accomplished to avoid additional water entrapment within the construction site?
- Have borrow materials that were used for embankments been compacted in layers as per specifications? Note: Proper placement and compaction are critical, perhaps the most critical phase of embankment construction.
- Have slopes been stabilized per specifications? Notes: (1) Slope stabilization will usually consist of soil erosion control matting, riprap protection, or a bioengineered slope protection. (2) The tops of the embankments are usually graveled to provide an all-weather travel surface.
- Have soil densities been verified by an authorized materials testing firm?
- Has the finished site grading been verified to meet the tolerances? Notes: (1) Grades should be specified so as to eliminate water ponding. (2) Final site grading should be verified before allowing topsoil or pavement operations to proceed.
- Do finished grades at building lines provide for positive drainage away from building walls, unless otherwise specified?

- Are all drainage trenches that are designed to carry water uniformly graded to their lowest point of discharge?

SOIL TREATMENT FOR TERMITES

- Have both the applicator's qualifications and materials to be used been approved.
- Is the mixing of materials and rate of application being checked periodically throughout the soil treatment operation to ensure that both comply with requirements of the specifications?
- Was all organic matter and debris that were in the areas to be treated removed prior to application of chemicals?

UTILITIES

Interruptions of utility service must be coordinated with all those affected. The line and grade of trenches should be checked before any pipe is laid and after the completion of each section. For most types of pipes, manufacturers have specified installation instructions that should be used; ensure that these have been followed.

- Have plans and specifications been thoroughly reviewed prior to the start of construction? Notes: (1) Plans and specifications should be reviewed for the location of existing utilities to avoid damage to them. (2) Lines and grades should be established and staked before any excavation or utility operations begin.
- Has each system been checked for conflicts at each point of crossing?
- Have all planned connections to existing utilities been verified and checked?
- Has all water line piping been laid straight between changes in alignment or direction and at uniform grade as indicated on the contract drawings and specifications?
- Is all water line piping being kept free of excavated material and other foreign substances?
- Has all water line piping been cut straight and true, leaving a smooth right angle cut?
- Is the contractor using a laser instrument or other approved methods to lay water piping, thereby ensuring compliance with contract invert requirements?
- Have thrust blocks been placed behind tees, bends, and hydrants? Notes: (1) Thrust pipes should be centered so that the thrust of water pressure will be exerted against the center of the block. (2) A bond breaker (tarpaper or polyethylene) should be placed between the pipe fitting and concrete.
- Have pressure or leak tests been performed prior to backfill placement over water pipe joints or fittings? Note: Testing requirements vary with the type of pipe and its intended purpose. Contract specifications usually detail the method of testing.
- Have potable water distribution lines been sterilized as (is usually) required by contract specifications? Notes: (1) Generally water lines are flushed with clean water to remove any mud or debris that may be in the pipe. Then the line's volume is calculated and a chemical disinfectant is introduced into it and retained for a designated period of time. The line is again flushed with clean

water until its discharge shows no further signs of the chemical disinfectant. (2) It is normal to have water samples taken and analyzed by an independent testing laboratory.

- Was drain line construction begun from the farthest downstream manhole, and is it proceeding "uphill"? Note: For bell and spigot pipes, the bell ends are usually required to point upstream; check specifications to verify this.
- Have drain line lengths between manholes been lamped (or inspected using laser instruments) prior to being placed into service? Note: A convenient method of doing this is to use a strong light that is placed in the downstream manhole and pointed into the upstream line. The inspector, using a mirror, can then check alignment in the upstream manhole.
- Have infiltration tests been conducted when required? Note: To conduct these tests, the ends of lines to be checked should be plugged in the upstream manholes and weirs placed in the downstream manhole. Infiltration should not exceed the amount specified. The weir must be cut to the shape of the manhole channel and caulked in place to prevent leakage.
- Have exfiltration tests been conducted when required? Note: Refer to specification on how to conduct exfiltration tests.

ROADS AND PARKING AREAS

Paving is specialized. Do not hesitate to request that an engineer with this specialized knowledge help with your inspections.

As work progresses from preparation, through to completion, you should continually refer back to the plans and specifications on matters such as grades, materials, and compaction requirements, and the location and nature of all utilities and drainage structures.

You should record all equipment that is brought on-site. Include type, make, condition, and presence of safety items; verify that safety items are operational.

- General requirements:
  - Have all road and parking area subgrades and embankments been tested to ensure proper compaction? Notes: (1) Subgrades are the foundation upon which the base courses and surface courses will be placed. It therefore goes without saying that subgrade compaction is as important as base course and surface course compaction. (2) Ensure that the geotechnical fabrics meet contract specifications and that they are installed according to manufacturer's recommendations.
  - Has the project area been properly cleared, grubbed, and stripped as per plans and specifications? Notes: (1) This should be done prior to the start of work. (2) Inspect to ensure the entire site has been properly prepared, for example, confirm the extent of cut and fill areas and the nature of soils that are left in-place.
  - Has the contractor laid-out the site as per plans and specifications? Note: Inspectors should familiarize themselves with required grades, drainage features, and all embedded items both existing and those to be installed.
  - Have all soil and compaction requirements in cut and fill areas been met?
  - Have the contractor's embankment construction and compaction methods been in accordance with contract requirements? Note: While

most clearing, grubbing, stripping, excavation, backfilling, and spoiling work was completed prior to the start of work, inspectors should continue to monitor these activities as construction progresses.

☐ Are the materials encountered as indicated on contract drawings? Note: If not, notify the general contractor GC or CM for appropriate action. Also, keep careful track of quantities as this work may result in extra costs.

☐ Are only suitable materials from excavations being used as fill material? Note: Pockets of soft, yielding, or otherwise unsuitable material should be replaced with suitable material.

☐ Is the contractor maintaining positive drainage as work progresses?

☐ Are surfaces in cut areas being compacted to the specified density?

☐ Do all subgrades meet grade and surface tolerances? Note: If subdrains are required, ensure that they are being installed as per plans and specifications.

☐ Are trenches backfilled with satisfactory materials that were compacted in specified layer thickness?

☐ Have required compaction tests been performed on subgrade? Notes: (1) Inspectors should require additional tests when there is any doubt. (2) Ensure that all test results are properly documented.

☐ Have offset reference stakes been set so that the position of utilities may be determined after installation?

☐ Have bank run[3] borrow pits been stripped of all unsatisfactory overburden? Notes: (1) Ensure "bank run" materials are stockpiled in cleaned and leveled areas. (2) If the material is plant processed, inspectors should make certain that processing, handling, and stockpiling methods will produce uniformly acceptable material. (3) Ensure that borrow pits are finished as specified.

☐ With regards to hauling and spreading, is spreading being carefully controlled to minimize segregation and to attain complete and uniform coverage?

☐ Are layer thicknesses as per plans and specifications?

☐ Are compaction operations as per specifications? Notes: (1) Be on the lookout for ruts and soft-yielding spots (pumping) as compaction progresses. Action must be taken to correct such weak spots by stabilization replacement of the unsatisfactory material. (2) Ensure that soil moisture content is as specified to ensure maximum compaction. (3) Check to ensure that the temperature and weather outlook is satisfactory for paving operations. (4) Specifications usually cover special procedures necessary for the soil type used.

☐ Are shoulders of specified width and are they placed and compacted at the edges of each layer of sub-base and base course?

• Bituminous prime coat[4]: An asphalt prime coat may be required prior to any additional surface treatment. Priming should penetrate the surface to plug

---

[3] As used here bank run refers to a bank, mound, dike, or the like, that has been raised to hold back water, carry a roadway, etc.

[4] A handy reference guide for paving and surfacing operations is US Army Field Manual, FM 5-436, *Paving and Surfacing Operations*.

capillary voids, coat and bond dust and loose mineral particles, and stabilize the surface and promote adhesion.

☐ Do the prime and tack coat bitumen materials comply with the specified viscosities or the approved submittals? Note: inspectors should familiarize themselves with the requirements for application and storage of the bitumen's.

☐ Has the rate of application of bitumen for both prime and tack coats been determined? Notes: (1) Inspectors should know the area measurement that will receive prime and tack coats. (2) Record quantities of bitumen that is used each day.

☐ Does all distribution equipment conform to the requirements of the job? Note: Check for proper heating and circulation of bitumen, control of spreading rate, and uniformity of application and measuring and indicating devices.

☐ Are all base areas and pavement courses clean and free of foreign material and water? Note: Ensure that the area to be primed or tacked is protected prior to and during paving operations.

☐ With **work in progress,** are weight bills and delivery tickets accompanying each delivery? Note: Verify that the required amount and approved bitumen is being applied.

☐ Is all equipment running smoothly and is the rate and temperature of bitumen application as specified and uniform? Note: Be prepared to stop operations if this is not so.

☐ Does the prime coat completely seal all surface voids without a surplus remaining on the surface after the curing period? Note: Take prompt corrective action in the event of unsatisfactory distribution.

☐ Has all bitumen been cured within the specified cure time? Note: Cure time is generally about 48 hours.

• Hot-mix asphaltic concrete and bituminous road mix surface course: Contractors can usually obtain prepared and mixed bituminous surfacing material from established plants; this is the assumption here. This may require inspectors to go off-site to ensure compliance with specifications.

Since we are not talking about large highway projects, there are two general methods of mixing aggregates with asphaltic materials directly on site, where the mixture is to be laid and compacted: blading and dragging, a method that is generally not permitted except on very small jobs, and traveling plant method.

☐ Have job mix formulas been furnished for approval? Notes: (1) The job mix should give the percentage of aggregate for each sieve size and the percentage of bitumen in the completed mixture. (2) The sources from which aggregate is used must be the same as the approved samples. (3) Aggregate gradation should be checked periodically to determine conformance with approved material gradation.

☐ Are surfaces dry when surfacing is laid? Notes: (1) Surface courses should not be placed when atmospheric temperature is below the specified temperature.

☐ Do lines and grades conform with the drawings and specifications?

□ Has the pavement mixture been completely mixed prior to spreading to determine whether mixing is complete? If the mix is not satisfactory, inform the contractor about corrections.

□ Is a mechanical spreader being used to spread the pavement mixture? Notes: (1) Spreading should be in accordance with specifications requirements. (2) Ensure that paving materials have been brought to their correct grades and that sections are as shown on contract drawings. (3) Immediately after any course is placed and before roller compaction is started, inspect the surface for any irregularities in alignment and grade. Look especially along the outside edges. Have necessary additions or removals of mixture been made before the rolling begins. At deep or irregular sections, intersections, turnouts, or driveways, where it is impractical to spread and finish the base using equipment, approved spreading equipment or acceptable hand methods should be used.

□ Are roller operations continuous after each layer has been spread? Notes: (1) After each layer has been satisfactorily spread, the surface should be rolled; a tack coat between layers may be required. (2) Rollers should provide equal compaction throughout? (3) Final rolling should be done using smooth–faced, power-driven rollers. (4) All power-driven tandem and three-wheel rollers should be in good condition, capable of reversing without backlash. They should be kept in continuous operation as nearly as practicable; all parts of the pavement should receive substantially equal compression. (5) Roller speed should not exceed three miles per hour and should be slow enough to avoid displacement of the surface course. (6) Any displacements occurring as a result of reversing the direction of the roller, or from any other cause, should be corrected immediately.

□ Is rolling being done in a longitudinal direction, beginning at the outer edges and working toward the center? Notes: (1) Each pass should overlap the prior pass by one-half the width of the roller. (2) All roller marks should be eliminated. This is generally a sign that no further consolidation is possible. (3) If the surface course adheres to the rollers, the rollers should be kept properly moistened, ensuring, however, that no excess water or oil is used.

□ Have the required number of samples been taken and tested?

□ Is the finished surface free from depressions that exceed 1/4-inch as measured with a 10-foot straight edge paralleling the centerline of the paved area, or at right angles to the centerline? Notes: (1) Test for conformity with the specified crown and grade immediately after initial rolling. Removing or adding materials should be accomplished to correct any variations in crown, grade, or smoothness; rolling should continue. (2) After the rolling is completed, the smoothness should again be checked and deviations should be corrected.

WATER WELL DRILLING

Usually, licenses issued to general contractors do not authorize well drilling; well drilling requires a separate license. So you should expect that a subcontractor will do this work. Given here is a short briefing on wells. Hopefully it will help you to communicate with your contractor.

Siting will not be in your choice; others will do this. You should, however, ensure that the location of drilling matches the contract drawings and location of local service controls.

Drilling rigs are of two general types: cable tool and rotary drill; some contracts specify which will be used. Cable tool rigs consist of a heavy drill bit that is connected to free-running wire cables. The cutting action is accomplished by the dropping of the bit and subsequent lifting by cable. Rotary drill bits consist of a set of wheel cutting devices attached to a rotating shaft. They use drilling mud as a lubricant and stabilizer to keep the hole open as drilling progresses. Some rotary rigs are equipped with a percussion hammer for driving the casing (similar to a cable tool rig).

The initial borehole may be drilled so as to accommodate the casing, or it may be started with a small hole and then reamed for the casing. This initial drilling is known as *spudding*[5] *in*. It is important that the hole be plumb throughout, so spudding in can't be faked, it must be plumb. Signs of wear or shiny sections on the cable or shaft are indications of being out of plumb. During this phase of the work, inspectors should continue to observe operations and record all observations and data obtained from the drilling.

Most well drillers use a casing to prevent caving during and after completion of a well. Casings keep out surface water that might contaminate the well. As a rule, the casing of each well should project ≥6 inches above the established grade at the well. However, most specifications require that the casing project ≥12 inches. A larger casing is generally used at the start of drilling operations, and is then reduced in diameter as the well drilling proceeds (domestic wells, 4 inches and 6 inches in diameter, normally use size casing with a well screen placed at the bottom of the hole.)

Materials are selected for strength and resistance in corrosion. Corrosion potential is highest closest to the surface of the ground where there is more moisture and air. Electrolytic corrosion can also occur when dissimilar metals are used in contact with each other. Some metals that may by themselves resist corrosion (e.g., bronze, brass, copper, aluminum) may corrode, or cause others to corrode, when placed in contact with a different type of metal. Rubber, plastic, or some other non-conductor should insulate different metals from each other. Care should also be taken in the selection of welding materials, as the weld connection is frequently the point where corrosion begins.

Casings must either be perforated or replaced by a well screen to allow water to enter the well. Perforations can be made in the field, if permitted by the specifications; otherwise, machine-cut perforated casings are available. Field perforating can be done before placement by punching or by cutting slots with an acetylene torch. In-place perforation can be made with a well knife or well perforator. Well screens that are available in varying designs are often inserted as casing sections. Well screens are particularly advantageous in sandy aquifers as here the screen opening can be selected to filter out sand exceeding a certain size. Where screens are necessary, inspectors should require samples of the water-bearing stratum and a copy of the well log and send them to a screen manufacturer for analysis and recommendations.

A gravel-packed well is one containing a gravel envelope surrounding the perforated portion of casing or well screen. The gravel increases the effective well diameter of the well, protects the casing, and acts to keep fine material out of the well. The type of formation and method of drilling determine the thickness of the gravel layer.

---

[5] Spudding: To begin drilling operations on a well, be it for water or oil.

Development is the essential operation that brings a well to maximum available capacity. Development unclogs the formation that occurs as a side effect of drilling. Every method of well drilling plugs off the pores of the water-bearing formation in varying degrees, and rotary drilling with mud fluid effectively seals off the bore hole. Development work stabilizes the sand formation around a screened well by removing the finer particles to increase the porosity of the formation. This also applies to gravel-packed wells.

The most common method for developing a well is surging and bailing. A steel cylinder (heavy bailer) with a one-way valve is moved up and down inside the casing or bore hole. On the down stroke, it forces water out of the well, and on the upstroke, it draws water and fines into the well. Surging and bailing continues until the amount of particles entering the well is at an acceptable level and the water is relatively clear.

Other methods of developing include surge plunger, valve type plunger, surging with air, jetting with air, jetting with water, pumping, and explosives. The well driller should know which method would give the best results.

Pump tests determine well output. They are also used to select permanent pump sizes, ensuring not to exceed any well's yield. The critical elements of a pump test are taking accurate measurements from the top of the well to the water level inside the casing. These measurements should be done before the test, during the test, and after pumping has ceased; accurate measurements of gallons per minute being pumped; and, that the pump test is continuous for the length of time specified. If the pump test is interrupted for any reason, the test should be repeated after the well has rested and the static water level returned to original elevation.

Grouting involves filling the space around the pipe, usually between the well casing and the borehole. If the well construction includes both an inner and outer casing, grout may be placed between the two casings in addition to sealing outside of the outer casing. Grout normally consists of a mixture of Portland cement and water, and the payment item is per bag of cement. Inspectors should closely observe the number of bags delivered on-site and collect empty bags on an interim basis for payment computations.

After both quantity and quality of the water seem to meet specifications, chemical treatment is used to clean out all the driller's mud, the well is disinfected and the water is analyzed per contract requirements. Then the well should be measured and capped. Temporary capping during the construction period should be required at all times when work is not active. If, for whatever reason a well is abandoned, it should be located topographically so that the exact site can be easily reestablished. Some states require that abandoned wells be fully grouted, thus permanently closed.

- Is the contractor furnishing the required type of rig? Notes: (1) Inspect the rig and all appurtenances to ensure that they have been well maintained. (2) Ensure that the driller's platform is able to carry the weight of the equipment.
- Has the contractor provided the specified gravel? Notes: (1) Gravel size and gradation are important if sand is to be held back at the outer edge of the pack. (2) Gravel should be installed to the depth required by the specifications. This is normally several feet above the top of the well screen.
- Is the water of an acceptable purity and clarity at the completion of development?

- Have you, the inspector, recorded the following information in the Daily Log during water well development:
  - ☐ Weather?
  - ☐ Start and end of the workday and length of time required for each particular operation, for example, obtaining specific formation samples?
  - ☐ Progress? Notes: (1) Record the following: the lineal feet and size of casing used; the well's depth and sample all changes of formations; all material classifications and their depths; water depths (if there is any gain in water, record the depth, run a bailing test, and obtain the rate of water increase and if the well is dry record this too); every time a well is to be left for a few hours, read the water level when leaving and again when work is resumed; the position and thickness of caving strata (pay particular attention when entering sandstone or gravel); driller's opinions of material, and so forth, along with you, the inspector's opinions. (2) Do not ignore as little as a 6-inches layer of sand that has only a trace of water. This may be enough water to develop a well from it, especially if a rotary drill is being used; test for quantity of water available in these cases before proceeding with deeper drilling.
  - ☐ Material deliveries?
  - ☐ All out of the ordinary events, for example, obstacles encountered, formations encountered (if known), and so forth?
  - ☐ All directives and/or recommendations made by the owner's representative?
- Has sampling been continuous, and in a manner that it represents the depth at which it was taken? Note: Each sample should be packaged and labeled properly?
- Has a boring log been kept for each hole?
- Are screens of the type specified?
- Are gravel packs of the type of materials, gradations, cleanliness, and so forth specified?
- Were screens inserted properly?
- Have all drilling sites been cleaned up before any contractor leaves an area?
- Has a complete report been written after each well is completed giving the final measurements of all pay items and furnishing test pumping and recovery records?

CONCRETE

In this section, commonly encountered concrete work is covered; some specialized procedures are covered in subsequent sections. Contract drawings and specifications for footings and slabs usually include the preparation of the subgrade prior to the pouring of concrete. It is also usual that specifications differ from job to job. Therefore, read the specifications and construction notes that include reinforcing steel locations and concrete thicknesses, elevations, and locations. The specifications/notes should also include a statement about the method of curing to be employed, along with weather conditions at placement time; study the plans and specifications carefully. You should take photographs to document construction.

- Have all materials and mix designs been approved prior to any pour?
- Has an independent testing laboratory that will make all required tests been named by the contractor? Notes: (1) Tests generally included are slump, air content, temperature, and curing and testing of test cylinders. You should verify that all required tests are being performed, including their number, curing time, and frequency. (2) If the specifications require the owner to take all tests, then, as the inspector you should become familiar with the correct procedures for all tests. Make sure that all required equipment is available prior to starting any concrete work.
- Are all w/c ratio[6] or slump tests within the specified bounds? Note: If the w/c ratio is exceeded, the load should be rejected and the contractor notified. The daily log should state the basis for rejection.
- Is the ready-mix concrete company providing delivery/load (batch) tickets to the contractor for each delivery? Notes: (1) You should receive a copy of each ticket before any concrete is placed. (2) Tickets should provide the following information:
  - Name of batch plant
  - Serial number of ticket
  - Date
  - Vehicle/license number
  - Name of purchaser/contractor
  - Project location/name
  - Specific class of the concrete, showing that it is in conformance with that specifications and that it has been approved. Note: The ticket should indicate that all ingredients are as previously certified.
  - Amount of concrete (cubic yards)
  - Time-loaded weights and types of ingredients (cement/aggregate/water)
  - Type, names, and amounts of admixtures
  - Additional information required by purchaser
  - Revolution counter reading at first addition of water
  - Amount of water added at site
- Is concrete being placed in a continuous manner and as rapidly as practical to avoid cold joints? Notes: (1) Placement should be conducted at such a rate that the concrete is at all times plastic and flows readily into space between reinforcing bars. (2) Concrete that has obtained its initial set, or has contained its water content for more than the allotted time, must not be deposited. (3) Concrete should not be dropped freely more than the specified distance and should be conveyed from the mixer to the place of final deposit by a method that will prevent the segregation or loss of the materials. (4) Concrete should be deposited as nearly as practical in its final position to avoid segregation due to re-handling or flowing. (5) Concrete placed in forms should be in layers of not more than the specified depth and each layer should be consolidated with the aid of vibrating equipment supplemented by hand spading, rodding, or tamping.

---

[6] The water–cement (w/c) ratio is the ratio of the weight of water to the weight of cement used in a concrete mix. A lower w/c ratio leads to higher strength and durability, but may make the mix more difficult to place, which is why some contractors try to get by with adding water. Placement difficulties can be resolved by using plasticizers or super-plasticizers.

- Are vibrators being used correctly? Notes: (1) Vibrators should not be used to transport concrete inside the forms. (2) A sufficient number of vibrators should be available to consolidate the concrete as required in accordance with the amount placed, so that no concrete will stand in the forms for more than 10 minutes before it is vibrated. (3) At least one spare vibrator should be maintained on the project at all times as a replacement. (4) The duration of vibration should be limited to that necessary to produce satisfactory consolidation without causing segregation. (5) Vibrators should not be placed between forms and the outer row of reinforcing. Vibrators should not ordinarily be permitted to touch forms.
- Is the contractor keeping the forms and shoring in place until the concrete members can safely support their weight and accompanying load?
- Is finishing being accomplished by the method specified? Note: The repair of defective areas and removal of fins, form marks, and holes should be done immediately upon removal of forms. Surfaces should be brought to their specified smoothness tolerances and rough areas and high spots should be ground smooth.
- Is all fresh concrete being maintained in a moist condition for the specified duration? Note: Fresh concrete must be protected from all damage including rain, too rapid drying, freezing, premature loading, and anything else that might affect its strength during the curing period. Therefore, curing should always be properly controlled and inspected. Freshly placed concrete should be inspected several times daily. Small tears and openings in the curing paper or polyethylene film should be covered with another piece of paper to prevent the loss of water through evaporation.
- Is backfilling and additional construction being kept away from fresh concrete for at least 7 days after placement or as specified?

LIGHTWEIGHT CONCRETE

- Is the proportioning, mixing, and placing of lightweight concrete being accomplished as per the supplier's directions? Note: This is important to obtain homogeneous concrete that has the characteristics that are set forth in the specifications.
- Is the surface of finished lightweight concrete being adequately protected from damage by heat, cold, rain, snow, direct rays of sun, and wind? Note: Lightweight concrete should be cured and allowed to dry suitably prior to the installation of roofing materials (or slabs on metal pans). This protection should remain until the roofing slabs have been installed.
- Is testing being conducted as required by specifications? Note: Inspectors should not hesitate to require additional testing if there is any doubt as to the quality of the lightweight concrete.
- Is all roller-compacted concrete (RCC)[7] and soil cement base course construction being accomplished with supervisors on site and workers who are sufficiently skilled to meet contract requirements?

---

[7] Roller compacted concrete (RCC): A mix of cement/fly ash, water, sand, aggregate and common additives, but containing much less water than more normal concrete; RCC is drier and essentially has no slump. The material is placed in a manner similar to paving, i.e., the material is delivered by dump trucks or conveyors, it is then spread by small bulldozers or modified asphalt pavers, and then it is compacted by vibratory rollers.

COLD WEATHER OPERATIONS[8]

Usually, the minimum requirements for cold weather concreting shall consist of the following:

- Are cold weather concreting operations being conducted in accordance with the American Concrete Institute (ACI) limits of ACI-306 and hot weather concreting operations in accordance with the limits of ACI-305? Note: Inspectors should become familiar with these standards.
- The ACI has developed the following chart that provides minimum concrete temperatures at the time of placement as a function of thickness;

	SECTION SIZE, MINIMUM DIMENSION			
Minimum concrete	12 inches	12–36 inches	36–72 inches	72 inches
Temperature as placed and maintained	55°F	50°F	45°F	40°F

- Is adequate protection from the weather being provided? Notes: (1) Adequate protection includes, if needed, the use of artificial heat, to prevent the temperature of the concrete from falling below the specified temperature for a period of 3 days when using type I cement and 2 days when using a set accelerator or type III cement. Specification may allow an alternate that allows a somewhat lower temperature for a longer period of time. Read the specifications.
- Where allowed, are type III[9] cement and non-chloride accelerators being used as part of a cold weather concreting plan? Note: Chloride accelerators (such as calcium chloride), fly ash, and ground-granulated blast-furnace slag chloride are usually not allowed to speed the hardening of concrete.
- Are loads, such as the placement of backfill against walls or the supporting of heavy equipment, not being applied until the concrete has been tested to have at least 75% of its design strength? Test results should be recorded.

HOT WEATHER OPERATIONS[10]

- Have subgrades, steel reinforcement, and form work been moistened prior to concrete placement?
- Have temporary wind breaks been erected? Note: Windbreaks limit wind velocities and provide sunshades to reduce concrete surface temperatures.
- Have aggregates and mixing water been cooled to reduce the mixture's initial temperature?
- Are concrete mixtures such that they allow for rapid placement and consolidation?

---

[8] References: ACI 318, Chapters 4 and 5; ACI 306.1-90; and referenced ASTM Standards.

[9] Type III concrete: A high, early strength cement. It is ground finer and reacts faster than Type I, so the early strength gains are greater but its ultimate strength is not higher than Type I.

[10] References: ACI Committee 305, "Hot-Weather Concreting" (ACI 305R-99), American Concrete Institute, Farmington Hills, MI; PCA—Design and Control 14th Edition, Chapter 13 (2002), Portland Cement Association, Skokie, IL; and NRMCA Publication #12, "CIP #12 Hot Weather Concreting" (2000), National Ready Mixed Concrete Association, Silver Spring, MD.

- Are concrete surfaces being protected with plastic sheeting or evaporation retarders during placement? Note: This should be done to maintain the initial moisture in the concrete mixture.
- Is staffing adequate to minimize the time required to place and finish the concrete? Notes: (1) Hot weather conditions substantially shortened the times to initial and final set. (2) Consider recommending fogging the placement area to raise the relative humidity and satisfy moisture demand of the ambient air. (3) Contractors should proceed with curing operations as soon as possible. (4) In extreme conditions, consider adjusting concreting operations to early mornings or night times.

REINFORCING BARS (REBAR)

The inspection of reinforcing bars in walls and slabs is fairly straightforward, until it isn't straightforward. Ironworkers often shift rebar to avoid obstructions, such as small openings, pipe sleeves, electrical outlets, and so forth. Inspectors need to decide if this is OK; it usually is acceptable when the total number of bars is not reduced. If not specified, Appendix C in the Concrete Reinforcing Steel Institute's (CRSI) *Manual of Standard Practice* provides first bar spacing. Problems may develop because both the formwork and rebar fabricators are given tolerances. When the formwork tolerance ends up being negative and the rebar manufacturer's tolerance ends up positive, a situation may arise where there is too little concrete coverage over the bars. Such conflicts may require resolution by the designer. The bending and re-bending of installed rebar may also be controversial; CRSI's EDR No. 12 can be used as a guide. If dowels are to be bent on site, the inspector should discuss the bending procedure with the contractor to assure that the bends conform to ACI 315. Where large-diameter bars are involved, some preheating may be recommended by the designer to avoid brittle failure.

- Are mill test or bar coating reports on site? Notes: (1) These reports should be made available to both contractor and government inspectors. (2) Bar mill markings should identify the producing mill, the type and grade of steel, and the bar size. (3) Contractors and owners may also want independent testing laboratory reports on samples taken at the fabricator's shop or at the jobsite as verification of the producer's mill test report. (4) Reports should state grade of steel, tensile properties, chemical composition (and carbon equivalent if rebar is to be welded), and spacing and height of deformations. These values should be compared with those in the applicable ASTM International[11] standard. (5) Where corrosion protection measures are specified, inspectors should verify the class of protection of the bars and the supports.
- Are approved drawings on site for field-placing personnel and inspectors?
- Is there a material shipping schedule for you to schedule inspections?
- Have potential problems been identified? Notes: (1) Detailing the rebar drawings is difficult and subject to error. Inspectors should review these drawings with the contractor to identify difficult-to-place details, lack of specifics, drawing discrepancies, detailing, and placing errors. If necessary, field changes should be approved before errors are made. (2) Length and

---

[11] ASTM International was formerly the American Society for Testing and Materials (ASTM), ASTM International is an organization that establishes material standards.

location of lap splices should be clearly specified in the contract documents and on the approved placing drawings. If mechanical connections are to be installed in lieu of lap splices, inspectors should request evidence that the architect or engineer has approved the use of the connection; literature that describes the recommended installation procedures should be provided.

- Has an agreement been reached regarding tolerances?[12]Notes: (1) Tolerances should be discussed to identify those that are critical, the method of measurement, and the basis for rejection or acceptance. (2) Specifications usually provide the standard tolerances to be followed by referencing ACI 117. Ultimately, the inspector should establish the range of acceptability. Any disagreement between contractor and inspector incompatible tolerances should be referred to the design professional for resolution. Concrete shall not be placed until the reinforcement has been inspected and approved by the on-site inspector.

  The following tolerances are commonly allowed in the placement of reinforcing bars:

  □ Where a 1½-inch clear distance is shown between reinforcing bars and forms, allowable clear distance is 1⅛–1½-inches.

  □ Where a 2-inch clear distance is shown between reinforcing bars and forms, allowable clear distance is 1⅝–2 inches.

  □ Where a 3-inch clear distance is shown between reinforcing bars and earth or forms, the allowable clear distance is 2½–3 inches. Over excavation that is backfilled with concrete is usually not counted toward clear distance.

  □ The maximum variation from indicated reinforcing bar spacing: 1/12th of indicated spacing, but no reduction in amount of bars specified.

  □ The ends of all reinforcing bars should be covered with at least 1½ inches of concrete.

  □ Unless otherwise indicated on the drawings, splices of reinforcing bars should provide a lap of not less than 30 diameters of the smaller bar but not less than 12 inches. Welding should not be used to splice bars. Tack welding or use of heat to bend reinforcing steel can change the steel chemistry that can have a negative effect on strength and brittleness. Welded wire fabric should be lapped at least one mesh width.

  □ The use of temporary supports is usually not allowed because the removal of the supports cannot be completed without causing the steel to be moved. The reinforcing steel is designed to have a minimum cover for corrosion protection and bond.

  □ The use of higher strength steel has made the lap splices generally longer than 30-bar diameters. Some older standard drawings assume the use of 40-grade steel; 30-bar-diameter lap splices are acceptable in those cases.

- Field inspections:

  □ Are bar diameters and shapes (if bent) as specified?

  □ Are bar lengths, spacing, embedment, and bearing on a wall or beam as specified? Notes: (1) Inspectors should count the total number of pieces and measure the slab bar spacing and check these figures against the

---

[12] ACI 117, the ACI Detailing Manual, and the Concrete Reinforcing Steel Institute (CRSI) Manual of Standard Practice tabulate fabricating tolerances for rebar.

approved placing drawings and the structural drawings. (2) Inspectors should also check beam longitudinal dimensions, column vertical dimensions, and stirrup and tie spacing.

☐ Are concrete precast blocks/chair heights such that the specified cover and clearances will be met? Notes: (1) It is important to check the chairs or standees supporting slab and mat top bars for height. (2) The entire mat and cages should be checked for stability; these can easily be displaced during concrete placement. (3) Normally, side supports are not provided unless called for in the contract documents.

☐ Are all bars/mesh reinforcing for slabs and footings supported on precast concrete blocks or chairs of an approved type? Notes: (1) Supports should be spaced at intervals that are established based on the size of the reinforcement used. (2) Reinforcing should be kept to the height specified above the underside of the slab or subgrade.

COMMON REINFORCED STEEL BARS (REBAR) SIZES

BAR SIZE, DESIGNATION	NOMINAL DIAMETER (INCHES)
#3	0.375 (3/8") OK
#4	0.500 (1/2")
#5	0.625 (5/8")
#6	0.750 (3/4")
#7	0.875 (7/8")
#8	1.000 (1.0")
#9	1.128 (~1-1/8")
#10	1.270 (~1-1/4")
#11	1.410 (~1-7/16")
#14	1.693 (~1-11/16")
#18	2.257 (~2-1/4")

Notes: (1) Forming components are usually attached after rebar placement. (2) If applicable, inspectors should request lapping requirements in writing.

☐ Are reinforcing bars tied together so as to form a rigid mat for footings, walls, and slabs? Notes: (1) Ironworkers usually tie a minimum number of rebar intersections. If the specifications are not precise about the number of tied intersections, inspectors should normally accept the work unless it is apparent the mats or cages of reinforcing steel will be displaced from their inspected position during concreting; the contractor is responsible for tying bars. (2) Only coated tie wire should be used to tie coated bars; tack welding of rebar-to-rebar should not be allowed.

☐ Have mechanical connections, welded splices, and other nonroutine and critical components and activities received the attention they require to ensure quality?

☐ Has dirt, grease, and other deleterious adhesions been removed before concrete placement? Notes: (1) Water-soluble cutting oils do not significantly affect bond. (2) Rebar that has a light coating of rust should not be rejected. (3) Inspectors should review the specifications regarding acceptance or rejection criteria of [slightly] damaged epoxy or damaged galvanized coating. Recommended touch-up procedures should be completed prior to acceptance.

FORMS[13]

- Were forms designed under the direct supervision of a professional structural engineer, and do they conform with the relevant codes? Notes: (1) When reviewing the design, inspectors should ensure that the designer considered, the rate of the pour, strength, stiffness, and that the form dimensions match the structural outline of the structure. (2) Consider also whether there are guardrails, toe boards, brick guards, ladders, and adequate access for workers.
- Are the forms constructed from acceptable materials and are they of the shape, form, line, and grades specified? Notes: (1) Inspectors should check materials for size, species, grade, condition, and damage.
- Are forms sufficiently braced to prevent deformation under load? Notes: (1) Items to check: width; height; plumb, inserts, ties, and such things as kicker panels, joints, and so forth. (2) Check to ensure that plywood lay-up is proper.
- Are forms constructed so as to prevent leakage at the joints (i.e., mortar leaks, not minor seepage of small amounts of free water)? Notes: (1) Inspectors should check forms for line, grade, and stability. (2) Forms that are to be reused should be coated with an approved form oil or equal. (3) After each use, forms should be thoroughly cleaned and checked for holes or roughness and repaired as needed. If repair is not possible, they should be discarded.
- Are all joints (expansion, contraction, and construction) located as shown on contract drawings or otherwise approved? Notes: (1) If joint locations are not shown on the drawings, inspectors should check with their supervisors to determine the appropriate locations prior to pouring concrete. (2) Inspectors should verify that water stops are firmly secured, in their correct locations, undamaged, and spliced properly. (3) The inspector should keep a record of the make, model, diameter, and strength of ties.
- Have the forms been inspected to ensure that all the embedded items are in place (includes inserts, reinforcing steel, and all other embedded items shown on drawings)? Note: This inspection should ensure that reinforcing steel is of the size and shape specified and that it is clean and properly supported to remain in place during the placement. Supports should be accurately placed and reinforcing securely tied at intersections.
- Are form ties sufficiently strong to prevent form deflection? Notes: (1) Designers must consider concreting rate and concrete stiffness and strength. Forms may look strong but deflection may occur. (2) Form ties are usually manufactured snap-off metal form ties that come with published stress values. (3) After form removal, the remaining portion of the ties should be recessed ¾ inch from the outer surface and the whole should not be larger than 1-inch diameter.

---

[13] Relevant references for formwork are: ACI 301, Structural Concrete for Buildings; ACI 318, Building Code Requirements for Reinforced Concrete; ACI 347, Recommended Practice for Concrete Formwork.

- Is the form release agent colorless so that it does not stain the concrete or absorb moisture? Note: Ensure that the release agent has been applied prior to pouring concrete.
- Are forms for structure walls kept up for at least 24 hours or more after concrete placement? Note: When forms are removed in less than 7 days, the concrete should be sprayed with a curing compound or be kept wet continuously as specified.
- Are forms removed in such a way as to prevent damage to the concrete? Forms should be removed before walls are backfilled.
- Are form ties being removed flush with or below the concrete surface? Note: Form ties that are removed to a depth of ≥½ inch should be patched with dry-pack mortar. It is common for the dry-pack mortar to consist of one part Portland cement and three parts sand, with just enough water to produce a workable consistency.
- Is all "honeycombed," damaged, or otherwise defective concrete being removed, the area wetted, and then filled with a dry-pack mortar? Note: Damaged or defective concrete should be removed and repaired so as to retain structural integrity of the member.

MASONRY

This section covers brick, concrete block, cinder block, glass block, tile, stone, and other masonry construction.

- Have samples of all materials been submitted for approval as per specifications?
- Have the required sample panels been erected and do the panels compare satisfactorily? Notes: (1) Check the masonry against the sample panels. The materials, workmanship, and finished appearance should be the same. (2) Check masonry dimensions against existing foundations and structural framing.
- Have all required tests been performed and have test results been submitted?
- Do materials that are on site match the approved samples for color, texture, and grade?
- Have ambient temperatures been at or above those specified?
- Are openings, partitions, and equipment locations as per plan?
- Are control joints located at the joints of openings?

CONTROL JOINTS LOCATED AT THE JOINTS OF OPENINGS

Notes: (1) Ensure that joint reinforcement does not pass through control joints. Reinforcement should be terminated 2 inches from control joints, except when reinforcement is used to accommodate diaphragm tension. Smooth dowels should be used across control joints to minimize the bond between grout and dowel. (2) Be sure that courses are installed level and plumb. (3) Masons should wait for initial set of mortar before tooling joints. (4) All tooling should be completed before quitting work for the day. (5) If units are moved after mortar takes initial set, inspectors should have them removed and replaced using fresh mortar. (6) Excess mortar should be removed from faces of units and joints before the mortar sets up.

STRUCTURAL STEEL

Structural steel is defined as the steel that supports the structure. Reinforcing bars that are used in concrete, metal siding, roofing, and miscellaneous items such as metal windows, stair railings, pressed steel doorframes, and metal trim are not considered to be structural steel.

- Has the contractor or fabricator provided all required structural steel shop drawings? Note: Contractors are responsible for errors in steel fabrication and improperly prepared materials. Note: Shop drawings for large and complicated assemblies are usually supplemented by erection drawings that show where each piece of fabricated steel is to be placed and to provide information on end markings so that pieces may be assembled correctly.
- Have all discrepancies between shop/erection drawings and the contract been resolved?
- Has the structural steel been inspected before erection for the following:
  - Size/Shape. Notes: (1) Check sizes and type of bolts, washers, and welds as well as hole diameters and location. (2) Check for beams that are made up of welded plates to ensure that they are not being substituted for a rolled beam.
  - Alignment and damage. Note: Straightening of bent or misaligned members should not be allowed.
  - New steel. Check to ensure that the furnished steel is new and has not been reworked in any way. Notes: (1) The steel should be free of rust. (2) Steel should be stored in a location that is away from construction traffic, it should also not be scattered over the site. (3) Steel members should be blocked off the ground to avoid corrosion and to aid inspection. For prolonged storage, the steel must be properly protected against the elements. (4) Care should be taken when handling steel members to prevent distortion or damage.
- Are base plates accurately located and level and does grouting fully cover the underside of the plates?
- Are frames plumbed and properly guyed? Note: This should be done prior to making final adjustments to setting. Ensure that all steel members are accurately fitted, leveled, and guyed before permanent connections are completed.
- Have all required welds been made, and are they accurately located and of specified sizes?

- Are all bolts, heads, and nuts resting squarely against the metal and are they drawn tight.
- Have brazed areas, abraded spots, and other damages to the shop coat been touched up with specified paint?
- Have contact surfaces at joints that are connected with high tensile bolts been painted/not painted as per specifications?
- Have both painted and unpainted steel been examined for loose mill scale or rust?

CARPENTRY AND MILLWORK

You should verify that all shop drawings have been approved prior to installation. All items that are delivered to the project should be checked against the approved samples. The questions in this section apply equally to light construction.

- With regards to general framing, does all lumber have required official-grade markings or has the contractor provided a certificate that covers the grade of lumber? Note: Regardless of grade markings, the presence of large or loose knots or other defects that would impair the lumber's strength/durability should be rejected.
- Is lumber being stored off the ground to ensure proper drainage, ventilation, and protection from weather? Note: Interior finish wood materials such as doors, floors, millwork, and so forth, should be in weather-tight, dry storage.
- Has treated lumber been inspected for compliance with the specification? Note: A certification of the treatment should be available.
- Have the exposed portions of treated lumber that was cut after treatment been brush coated with a compatible preservative?
- Is all framing close fitting and rigidly secured in place?
- Are cuts, notches, or borings in structural members for the passage of utilities done in accordance with plans, specifications, and the building code?
- Have damaged framing members been reinforced or replaced?
- Are all joints in framing timbers and girders neatly made and do they provide a solid bearing over the entire area of the joint? Note: Joists and beams that support weight should not be reduced in bearing area by notching or cutting away their undersides without an engineer's approval.
- Is care being taken to protect finish woodwork from being damaged during installation?
- With regards to partitions and walls, are anchors for plates sized and spaced as per plans and specifications?
- Are headers over openings in compliance with specifications?
- Do all bearing partitions have double top plates?
- Do exterior walls have diagonal bracing as required by specifications?
- Is all blocking for support of fixtures of ample size, closely fitted, and rigidly secured in place?
- Are studs doubled at openings and tripled at corners?
- With regards to floors, do joists have at least a 4-inch bearing at each end?

- Are the ends of framing joists for masonry angled for "self-releasing?" Note: This is an often-ignored detail. In case fire burns through a joist and causes it to fall, this will preclude the supporting wall from being broken or tilted over. Full bearing should be provided at the bottom of the joist.

"SELF-RELEASING" FRAMING JOIST

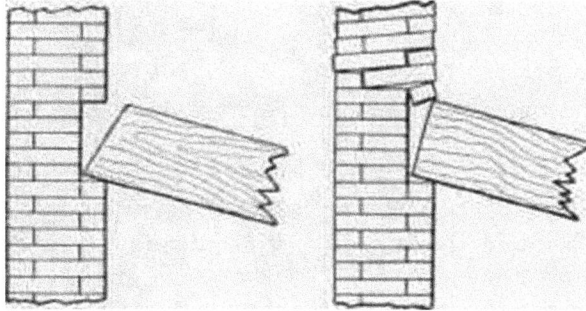

- Do floor joists rest on smooth level surfaces? Notes: (1) If foundation sills are not required, floor joists can rest directly upon the concrete or masonry foundation. Unfortunately, it is difficult to obtain a smooth, level surface on which to rest the joists, and it may be necessary to use shims. Where shims are necessary, their use must be in accordance with specifications.
- Are foundation sills level when placed on concrete or masonry foundations? Notes: (1) It is good practice to spread a bed of mortar on the foundation and lay the sill on it, tapping the sill to obtain even bearing throughout its length. (2) Anchor bolts and nuts should not be tightened until the mortar has set. (3) Sill sealer should be used per specifications for a watertight seal.
- Has the contractor constructed the framing, that is, joists, girders, beams, headers, joist hangers, notching, borings, anchors, ties, and so forth, exactly as per plans? Note: There are many examples of building failures due to construction not having been accomplished exactly as per plans. Inspectors should familiarize themselves with the details of the framing plans and have the plans available for constant checking as the work progresses.
- With regards to roofs, are the plates on masonry walls set level in beds of fresh mortar and secured with anchor bolts? Notes: (1) The nuts for anchor bolts should riot be pulled down tight until the mortar has set. (2) Check the drawings to determine whether wall plates should be single or double.
- Has blocking been provided between rafters as required to form nailers[14] for roof sheathing? Note: The specifications may require that hip and valley rafters be secured to wall plates with clip angles.
- With regards to bridging, has the top of the joist been brought to a straight line before placing and nailing the bridging? Notes: (1) Bridging can be cut

---

[14] As used here a "nailer" is a piece or pieces of dimensional lumber and/or plywood secured to the structural deck or walls that provide a receiving medium for the fasteners used to attach membrane or flashing.

and placed by nailing the top end to joist. Nailing of bottom ends of bridging should be deferred until after placement of subfloor and, if possible, after placement of finished floor to allow tops of joists to be drawn into better alignment. (2) Bridging strips should have ends accurately level cut to allow firm contact with sides of joists. (3) To prevent a sliding movement of metal bridging, nails should be correctly sized to fit the punched holes.

- With regards to wall sheathing, are vertical joints between plywood sheets offset so that the same stud does not carry the joint in succeeding rows of sheathing?
- Has waterproof sheathing paper been installed as required by specifications/plans.
- Are metal ties installed as required?
- Is plywood sheathing the correct thickness?
- Has fiberboard sheathing been installed with the required allowance for expansion at ends and edges, except at opening frames?
- Has fiberboard and gypsum sheathing been applied horizontally with joints staggered? Note: Ensure that the correct nails are used when mounting fiberboard and gypsum sheathing.
- With regards to roof sheathing, is the sheathing of the type and thickness specified?
- Has plywood sheathing been applied with the grain of the face plies across the rafters? Notes: (1) Ensure that nails are correctly spaced and fastened. (2) Specifications may require that exposed sheathing boards or planks, for example, at eaves be V-jointed, dressed, and matched.
- With regards to subflooring, are the butt joints of wood subflooring staggered and over supports, with ends cut parallel to joists and with adhesive used as per specifications?
- Has the top of the subflooring been inspected for a true, even plane?
- Has plywood subflooring been installed with the grain of face plies across joists? Note: Solid backing should be provided under edges at right angles to joists.
- With regards to framing, do stair stringers have solid bearing at both the top and bottom back edges of the stringers.
- With regards to window frames, are the frames constructed as specified and detailed.
- Are frames that are set in masonry wall braced so that the mason cannot knock them out of position when laying the brick or other masonry?
- Window units plumb and level without warp or rack of frames? Note: Double-hung windows should glide freely without binding.
- With regards to doors, have top and bottom edges of exterior wood doors been given specified coats of appropriate finish? Notes: (1) Look especially at the bottom of the doors that often go unpainted. This may lead to inordinate expansion and contraction due to changes in moisture. (2) If the top and/or bottom edges are cut at the job site, two coats of finish should be applied immediately.
- With regards to interior finish trim, are moldings mitered at corners and coped at angles?

- Are shoe moldings nailed to flooring and not to the base mold, except where finish floors are other than wood? Note: In this case, the shoe mold shall be nailed to the base.
- Are finish floors perpendicular to floor joists? Notes: (1) If subflooring has become warped or loosened, it should be re-nailed and completely tight before placing the finish floor. (2) Cross-joints in wood floors should be well distributed, and unless the flooring is end-matched and laid on a subfloor, joints should be made over solid bearings. (3) If the flooring drives up with difficulty, a wood block or piece of 2 × 4 inches should be used against the tongue to prevent damage. (4) Only flooring nails that are approved for installation should be used for nailing floors. (5) Nails used on tongue–and–groove flooring should be driven at an angle of 45° or more to obtain as much penetration in the joists or blocking as possible. (6) The ends of the flooring should be cut at a slight under bevel to permit the top of the board ends to form a tight joint.
- Have mill work manufacturers (millwork is generally produced in a shop or mill) submitted required shop drawings? Note: Upon delivery of the units, inspect them against the previously approved shop drawings, giving special attention to:
  - Types of materials and methods of jointing
  - Bracing at corners
  - Thicknesses of materials
  - Proper fit of doors
  - Drawer guides
  - Joints between counters and splashbacks
  - Finishes or sealers and primers.
- Is millwork being stored on site in weather-tight, dry facilities?

THERMAL AND MOISTURE PROTECTION

In general, this section covers installing roofing, waterproofing, thermal protection, exterior siding, fire stopping, and smoke protection and sealant systems.

- Do the specifications require that any "system" such as a roofing system to be of a single manufacturer? Note: This rule is to ensure chemical compatibility. Included should be the manufacturer's recommended accessories and adhesives. If not the case this should be brought to the designer's attention.
- Have design criteria, data, cost and lifecycle information, and so forth, been documented for all specialty roofing systems and LEED compliant systems?
- Are all materials for thermal and moisture protection systems stored in safe dry environments? Note: Wet or damaged materials should be rejected.
- Do all roof designs identify each item requiring fireproofing, the code required rating, and the proposed method of fireproofing? Note: (1) Cut sheets and the UL or Factory Mutual (FM) number, rating, and construction method should be part of the construction documents. (2) Manufacturer's data should be provided for all specialty fire stopping or fire-blocking materials, including UL or FM classification. (3) Installation of these products may require an independent agency inspection.

MEMBRANE ROOFING (BALLASTED AND GLUED)

Installed roofs tend to be neglected until they start leaking. You should have a full understanding and knowledge of the roofing specifications and details, including those of the manufacturer.

- Do all delivered materials comply with the specifications?
- Are materials being stacked and stored to prevent damage and to afford protection from the weather?
- Have all projections, protruding nails, surface irregularities, holes, and surface voids been corrected prior to application of roofing?
- Are all materials dry and free from moisture absorption prior to installation? Notes: (1) Ensure that the roof is clean prior to the start of work. (2) Do not apply roofing materials to damp, frozen, or dirty deck surfaces; adhere to specified temperature requirements.
- Have the locations of all penetrations and types of penetrations and perimeter been verified?
- Are cant strips uniform and smooth?
- Is decking properly supported and secure, and are all penetrations solidly set?
- Have roofing materials been distributed to avoid the deck from being overloaded?
- Have all tapered insulation, roofing boards, and vapor barriers been installed per manufacturer's instructions? Note: Materials should lie down smoothly and should fit neatly at roof breaks, around the perimeter, and at protrusions through the roof.
- Has membrane roofing been installed as per the manufacturer's written instructions? Note: Special attention should be given to cleaning, overlaps, bubbles, wrinkles, and end and edge joints.
- Are watertight penetrations and parapets sealed with approved flashing membrane material? Notes: (1) Ensure that water does not flow beneath completed sections of roofs. (2) Edges should be sealed whenever weather is threatening.
- Has traffic surfacing been placed as per plans?
- Has ballast been placed and distributed evenly to the thicknesses noted on drawings?

ASPHALT SHINGLE ROOFS

Usually the specifications describe the method of application, quality of materials, and quantity of materials or refer to manufacturers' recommendations. Inspectors need to study the plans and specifications to learn what surfaces and types of roofing are to be installed on the individual buildings. When materials are delivered to the project site, inspectors should verify that the materials are in accordance with previously approved samples.

- Are the roofing materials that have been received protected to prevent damage during storage? Note: A common standard for asphalt shingles is that they are wind-resistant type #280 or greater and that their fire resistant-rating is UL Class A.

- Has all bonding material been delivered in sealed containers that have the manufacturers' original labels?
- Have all projections, protruding nails, surface irregularities, loose boards, and large cracks been corrected so as not to damage shingles?
- Do the starter strips for mineral-surfaced, asphalt, strip-shingle roofing project one-half inch beyond the eave line to form a drip? Note: Check plans to determine if a drip strip is specified.
- Is the contractor ensuring that the roofing at valleys, hips, ridges, and flashing is water-tight and secure against high winds?
- Do sheet metal shingle flashing installations for all asphalt shingle roofs conform to the Sheet Metal and Air Conditioning Contractors National Association, Inc.'s Architectural Sheet Metal Manual?
- Is the contractor adhering to the minimum air temperature requirements?

SHEET METAL ROOFING

Commercial metal roofing consists of interlocking rippled metal sheets that are fastened directly to the roof. The panels are fastened through the panels with rubber gasket fasteners. Much like standing seam metal roofing (below), commercial metal is fire resistant and capable of withstanding a wide range of climates. When installed properly, sheet metal roofing may last up to 50 years.

- Have all required samples been submitted and approved? Notes: (1) Usually, specifications require that samples of items to be installed be submitted for approval. All materials that are received at the job site should be inspected to ensure conformance with the previously approved samples. (2) Materials that are fabricated should be checked against shop drawings, contract drawings, and specifications. (3) Be aware that the required gauge for non-ferrous sheet metal and ferrous (iron) sheet metal are of different thickness for the same gauge. When inspecting use the proper measuring gauges.
- Have screens for ventilation been inspected upon delivery to job site? Note: Check to ensure that requirements as to mesh, type of material, and frame construction are as specified.
- Is the soldering flux that is intended for use as specified?
- Has the entire project been evaluated to ensure that galvanic action may not occur? Notes: (1) Some common galvanic action errors are copper and aluminum flashings in contact with each other, or with ferrous (iron) material; copper and aluminum flashings nailed with ferrous nails; and aluminum and ferrous equipment bases set on copper flashings. Nails, screws, and bolts that are used for installing and fastening sheet metal should be of types that are best suited for the intended purpose and of a composition that will not support galvanic action. Where different metals do come together, the metals should be insulated from each other; it is important to ensure that the insulation will stand up to the weather and UV. (2) Typical examples to watch for are:
- Have all flashings been carefully inspected? Notes: (1) Ensure that the type of material specified, gauge, weight, and width, are as required for the various

Metal corroding	Contact metal	Magnesim & alloys	Zinc & alloys	Aluminium & alloys	Cadmium	Steel-carbon	Cast iron	Stainless steels	Lead, tin and alloys	Nickel	Brasses, nickel silvers	Copper	Bronzes, cupro-nickels	Nickels copper alloys	Nickels-chrome-mo alloys titanium, silver, graphite graphite, gold, platinum
Magnesim & alloys			X	X	X	X	X	X	X	X	X	X	X	X	X
Zinc & alloys				X	X	X	X	X	X	X	X	X	X	X	X
Aluminium & alloys					X	X	X	X	X	X	X	X	X	X	X
Cadmium						X	X	X	X	X	X	X	X	X	X
Steel-carbon					X		X	X	X	X	X	X	X	X	X
Cast iron						X		X	X	X	X	X	X	X	X
Stainless steels						X	X		X	X	X	X	X	X	X
Lead, tin and alloys								X		X	X	X	X	X	X
Nickel										X	X	X	X	X	X
Brasses, nickel silvers						X	X	X	X			X	X	X	X
Copper								X	X	X			X	X	X
Bronzes, cupro-nickels														X	X
Nickels copper alloys															X
Nickels-chrome-mo alloys titanium, silver, graphite graphite, gold, platinum															

X = Glavanic corrosin risk

types of flashings. (2) Every joint and seam should be thoroughly checked to ensure that no leaks would occur. (3) Verify that base flashing is installed at the edge of roofing and intersections of vertical or similar surfaces. (4) Ensure that cap flashing or counter-flashing is built into vertical walls and they turn down over base flashings for the distance required by the specifications.[15]

- Have gutters been firmly attached with screws or bolts that are spaced as specified and are they sloped throughout their length in the direction of flow? Note: Gutter joints should be inspected as thoroughly as other sheet metal joints.
- Have gravel stops been lapped and their joints soldered? Note: Check joints for insulation wherever dissimilar metals have been used.
- Are louvers rigid and vibration free?
- Do louvers have the correct type and color of paint?
- Are the edges of louver blades folded or beaded for strength?
- Have insect and bird screens been installed as per the specifications.

STANDING SEAM METAL ROOFS

Standing seam metal roofing is constructed of interlocking metal panels that run from the ridge of the roof to the eave. The seams of the two panels are raised above the surface to allow the water to run off without getting between the panels. The seams

---

[15] A water-resistant material that may be metal or membrane that extends through a wall and its cavities is called a "through-wall" flashing. Through-wall flashing is positioned to direct water entering the top of the wall or cavity to the exterior, usually through weep holes.

are fastened to the roof using one of two methods: (1) Hidden anchors are located on the raised portion of the panel. The adjacent panel overlaps the raised edge hiding the fastener. (2) Exposed fasteners are installed through the panel directly into the roof sheathing.

Standing seam metal roofs can withstand a wide range of climates. They come in a wide range of colors and styles, helping shed its commercial image. The vertical lines of a standing seam metal roof can function as an architectural design element. Standing seam metal roofs highlight the valleys, peaks, and gables of the roof.

Usually, specifications require that samples be submitted for approval. All materials that are received at the job site should be inspected to ensure conformance with the previously approved samples. Check thickness of metal panels with a sheet metal gauge.

- Are metal panels being stored in a clean, dry area or are they covered and sloped for drainage as necessary? Note: If a removable plastic film protects the panels, this film should not be removed until installation.
- Are metal panels protected from abuse by construction traffic and from contamination by corrosive or staining materials?
- Are stored materials and unfinished work being secured against wind damage?
- Are metal panels being installed on a substrate and/or subframe that has been installed and aligned to acceptable tolerances as recommended by the panel manufacturer?
- Has the contractor's provided walk boards in areas of heavy traffic or other measures taken to prevent damage from construction crews?
- Is all work being installed in accordance with approved shop details under direct supervision of an experienced sheet metal craftsman?
- Do attachments and joints allow for expansion and contraction from temperature changes without distortion or elongation of fastening holes?
- Have flashings been installed in strict accordance with recommended practices?
- Are panels caulked, sealed, and fastened to provide a complete weather-tight installation?
- Have all standing seam roof panels been mechanically seamed on the roof with a seaming tool? Note: The use of "snap lock" or friction-style seaming should not be allowed.
- Is excess scrap and debris being removed from the working surface and surrounding area as required or at least on a daily basis?
- Is the contractor touching up areas as required with the manufacturer's touch-up paint?
- Are panels free of stains and scratches? Note: The contractor should wash panel surfaces if necessary.

MEMBRANE ROOFING[16]

- Does the roof design provide ¼-inch pitch where possible? Note: In no case should the pitch be less than ⅛-inch per foot.
- Is all roofing insulation chemically compliant with the membrane and adhesives?

---

[16] Inspectors should have the right to perform test cuts of all new roofing to verify that the installation is per specifications. Contractors should be responsible for repair of the test cuts.

- Do membrane roof systems run up and over parapets (beneath copings/caps)?
- Are membrane roofing seams ≥5-feet from drains?
- Are piping supports of compatible composite material and ≥1-inch wider than the pipe by ≥4-inches long with piping securely fastened to the supports? Note: Supports should not be fastened to the roof, but allowed to "float."
- Have walking pads been installed on routes leading to all roof-top equipment?
- Have roof hatches and equipment been placed ≥10 feet from roof edges? Note: If not, safety rails are required.
- When adjacent roof areas sheet drain to lower roof areas, has a gutter system been installed?
- Are all through wall penetrations seamless or soldered joint metal box over tight membrane, fully flashed? Note: Collection boxes, gutters, and splash blocks should be provided, where appropriate.

INSULATION MATERIALS

Insulation materials are manufactured in many forms and types that vary in thermal properties for specific job requirements. While this is primarily a design issue, inspectors should attempt to stay current. For example, new energy code requirements such as the 2009 International Energy Conservation Code for residential buildings and ASHRAE 90.1-2007 for commercial buildings (as well as the even stricter 2012 IECC requirements that some communities may soon implement) are driving new methods for adding additional insulation to the building envelope.

- Is stored insulation protected from the elements?
- Is all insulation used on the project of the type, thickness, and quality as per specifications?
- Does all insulation material, including facings such as vapor barrier and breather paper that is installed within floor–ceiling assemblies, walls, crawl spaces or attics, have the flame spread rating that is specified? Notes: (1) Common specifications are a flame spread rating of ≤25 and a smoke density ≤450. (2) An exception may be made when such materials are installed in concealed spaces of types III, IV, and V construction. Here the flame spread and smoke developed limitations may not apply to facing provided that the facing is installed in substantial contact with the unexposed surface of the ceiling floors or wall finish.
- Have all floor sill plates and exterior openings been sealed? Note: Make sure all surfaces are clean prior to sealing.
- Has insulation been secured in place to prevent material from coming in contact with vents and other potentially hazardous appurtenances? Notes: (1) Check all clearances from vents, flues, combustible air chases, ducts, and ventilation. (2) Nonrated lights require ≥3-inch clearance on each side and above (see Electrical Rough-in).
- Have all pipe penetrations in floors been filled with insulation?

STUCCO

Stucco is a term that is applied to exterior plastering. Generally, it is applied in three coats: scratch coat, brown coat, and thin finish coat. It is important to cure each layer properly to develop maximum strength and density.

- Is one brand of material being used throughout a single structure? Note: If more than one brand of material is used, you will in all likelihood end-up with differences in finished colors and textures.
- Are fibers or hair free of dust and dirt? Note: Each fiber or hair serves as a bit of reinforcing, so when dust and dirt are present the mortar does not adhere properly to the fiber and hair.
- Is the grading of the sand as per specifications? Note: Finished coats may require somewhat finer aggregate. However, excessive fineness is a principal cause of cracking and crazing.
- Is all metal reinforcement straight, without buckles and with staggered joints? Notes: (1) The long dimension of metal reinforcing must be across the supports, to remove most of the sag. (2) Reinforcement that is applied to wood supports should be attached with nails or staples driven through furring sleeves (spacers) to provide space for mechanical keys behind the reinforcing. (A rough texture creates a good mechanical key for stucco, while a sand finish or smooth finish limits the mechanical key.). Expanding metal lath that is mounted on vertical supports should be placed with the middle web sloping in toward the vertical supports. This causes the stucco to slide in and form a good key.
- Is the contractor carefully measuring cement and aggregate for each batch of stucco to assure uniformity? Notes: (1) No water should be added to the cement and aggregate until they are thoroughly dry mixed. (2) Mechanical mixers, mixing boxes, and tools should be cleaned after each batch to prevent the use of stucco that has already taken its initial set; re-tempering should not be permitted. (3) Proportions are generally well defined in specifications and should be adhered to.
- Are surfaces that are to receive stucco clean and free from dust, dirt, oil, and so forth?
- Have masonry surfaces that are to receive stucco been uniformly dampened? Note: This prevents masonry from sucking moisture too quickly from the stucco.
- Have the scratch coats been applied with sufficient pressure to form good keying?
- Have the scratch coats been well scratched to provide a bond for the brown coats?
- Have the scratch coats been properly damp cured, as per specifications? Note: Each coat of stucco should be moistened as soon as it has set sufficiently so as not to be damaged.
- Have the surface of the scratch coats been evenly dampened with fog sprays to obtain uniform suction prior to the application of the brown coats? Note: The interval between each brown coat and finish coat should be per specifications.
- Have the brown coats been dampened thoroughly and evenly before applying the finish coats? Note: A brush should not be used for dampening.
- Is the ambient temperature above 40°F during stuccoing? Note: Stucco that is frozen and thawed will not cure properly.
- Is the stucco being protected from direct sunlight once the finished coat has taken its initial set? Notes: (1) Once the curing process has started, the stucco should be kept moist by spraying at intervals for the period required by specifications. (2) Do not use membrane-curing compounds for curing stucco.

DOORS AND WINDOWS

Doors and windows are usually taken for granted. But it is hard to imagine any part of a building that receives more use and abuse than these openings between the interior and exterior of a building; this is especially so of doors. They are also responsible for large amounts of heat loss and gain, depending on the weather. Therefore, your inspection of doors and windows is very important. If they are not installed properly, they will quickly go to ruin and will have to be replaced.

STEEL DOORS AND FRAMES

- Have shop drawings been submitted and approved prior to the installation of metal doors and frames? Notes: (1) A contractor should verify all measurements, as they are responsible for dimensions, fittings, and the proper attachment of items directly connected to the door installation. (2) Check the thickness of metal and all other features that may affect the strength of the door, for example, construction details, method of assembling sections, location, and installation of hardware, the size, shape, and thickness of all specified materials, joints, and connections. (3) The finished surfaces should be checked against the specification requirements. (4) Doors and frames should be properly marked with door opening mark numbers that correspond to the schedule.
- Have all steel doors been delivered with corrugated edge protection? Note: Doors should be palletized to provide protection during delivery.
- After delivery, have doors been properly stored? Notes: (1) Doors should be stored under cover so as to prevent rust and damage. (2) Do not store doors in non-vented or canvas shelters where high humidity can occur. (3) Provide ¼-inch space between stacked doors.
- Have the doors and frames been erected in accordance with the details contained on the shop drawings and on the contract drawings? Notes: (1) Ensure that hardware, weather stripping, the grouting of doorframes, and louvers are as per shop drawings, contract drawings and specifications. (2) Ensure that frames are properly supported and that doors swing freely.

FIRE DOORS[17]

Since we are talking about inspections that are taking place prior to issuance of a Certificate of Occupancy (CoFO), the building code still matters. But once the CofO is issued, the building code issue is closed and the Fire Code or Life Safety Code goes into effect for operation and maintenance of the facility.

- Have the fire door assemblies been evaluated against the following criteria:
  - □ There are no holes or breaks in doors or frames.
  - □ All glazing and glass kit/glass beads are intact and securely fastened.
  - □ All doors, frames, and hardware are in proper working order.
  - □ There are no missing or broken parts.
  - □ All door clearances are within allowable limits.
  - □ All door closer/spring hinges are operational and doors are self-closing.

---

[17] Primary references for this fire doors section are *Fire Door Assembly Inspection Checklists*, by Ingersoll Rand Security Technologies of New England and NFPA 80 2013, *Standard for Fire Doors*.

- All coordinators[18] ensure that door leaves close in proper sequence (pairs only).
- All doors are self-latching in the closed position.
- No openings are equipped with auxiliary hardware items that interfere with operation.
- No field modifications have been performed that void the label.
- Where required, are all gaskets and edge seals present, continuous, and of the proper type for a fire door.

GLASS AND GLAZING

You should become familiar with the types of glass that are being used on the project and where the glass will be placed. Ensure that each type, for example tempered glass, is proper for the place used and that it has been approved before use. Materials should be installed per manufacturer's recommendations and contract specifications.

- Is all prism glass being set with the prisms on the room side of sashes, so the prisms will reflect light into the room?
- Is all obscure glass (generally used for glazing doors and sashes of windows in toilets, baths, and dressing rooms) being set with their smooth surfaces to the exterior and with the surface design in one direction (unless otherwise required by specifications)?
- Is all glass and mirrors factory labeled on each pane? Note: Labels should not be removed until prior to being installed and inspected.

HARDWARE

In a large, multistoried building, hardware can become a nightmare. Keying schemes can quickly become confusing, as people will want their spaces to be secure, yet will also want custodians and other maintenance personnel to have access; the major hardware firms can usually be called upon to assist in locking schemes. Other fire and safety regulations may battle it out for preeminence. Good hardware is expensive and designers faced with funding issues may look at hardware as a place to cut costs, so hardware of lesser quality may be chosen, an unwise decision. Final acceptance of the hardware contract should require a demonstration that the hardware performs satisfactorily.

- Have all delivered hardware been compared against the hardware schedule? Note: Inspectors should insist that the hardware schedules for windows and doors be detailed and submitted early enough for a complete review.
- Are hinge pins plumb through all hinges on all doors? Note: All hinged, pivoted, sliding, or otherwise movable hardware should work free and easily.
- Do doors have the correct number of hinges? Note: Be sure that doors that require three hinges do, indeed, have three hinges.

---

[18] Coordinators ensure that the leaves of a pair close in the proper sequence. There are two common types of coordinators: gravity coordinators that are mounted on the frame face and bar-type coordinators that are mounted on the underside of the frame head. If both leaves of the pair are opened and the active leaf begins to close first, the coordinator holds the active leaf open a few degrees to allow the inactive leaf to close first. When the inactive leaf is in the closed position, it engages the release mechanism on the coordinator that then allows the active leaf to close.

- Do screen and storm doors clear door closers? Determine the number of keys required and have all keys tagged as to location.
- Are doors weather tight?
- Are handicapped entrances equipped with power units/controls and are they located properly?

INTERIOR FINISHES

This section covers interior exposed surfaces of a building, such as wood, plaster, and brick, or applied materials such as paint and wallpaper, thereby making everyone an expert.

CERAMIC AND QUARRY TILE

You should verify that all materials that are needed for installation of the tile have been approved; compare all material delivered with approved samples and literature. Check especially for size, color, and pattern of furnished tile. Regarding the installation of floor tile, inspectors should ensure that the walls are installed before floors are installed. Check layout, preparation of setting bed, and bed mix. Do not allow the contractor to place excessive setting bed ahead of tiling operation, since tile cannot be placed on material that has obtained initial set.

- Does the preparation for placement of tile comply with specifications? Note: Check location and anchorage of studs and furring; the scratch-coat and float-coat operations; the composition and workability of the mix used; and the application and curing.
- Is tile set straight, level, perpendicular and are joints uniform? Note: Ensure that all tiles are undamaged and set firmly in place.
- Have all accessories been properly secured?
- Is the grout properly tooled, cleaned, and cured?

RESILIENT FLOOR COVERING/TILE

Usually, contractors must submit samples and obtain approval of all materials needed for installation of the floor. Inspectors should compare the materials that are delivered to the site against the approved samples and the contract requirements. Inspectors should familiarize themselves with the placement pattern/design.

- Is all delivered tile as specified: size, thickness, color, and square edges? Note: Tile containers should not be opened prior to delivery to assure that the tile is of the grade and color previously approved. This also applies to containers holding the cementing materials for tile application.
- Have subfloors been vacuumed clean, free from dust, dry, and smooth before tiling begins? Notes: (1) To secure a good bond between the resilient floor covering/tile and subfloor, the subfloor must be clean. (2) Irregularities of the subfloor will tend to show through the finished tile; indented surface defects should be filled with plastic material that is manufactured for this purpose.
- Have materials been stored at above the minimum temperature and period specified before installation? Note: Similar precautions should be taken in all spaces where tile is to be installed.

- Where felt is required, are all edges butted, not lapped, and are they carefully cut to fit around vertical surfaces? Note: The felt should always be rolled with a linoleum-type roller of specified weight.
- Is sufficient time being allowed after application of cement for it to dry or set up? Note: The type of cement, temperature, humidity, and ventilation in the area governs the time required for cement to set up. Placing of tile on cement too early or too late results in incomplete bonding of tile.
- Have finished floors been thoroughly cleaned and protected from traffic? Note: To prevent damage to the tile, cleaning should not be done with a solvent-type cleaner.

ACOUSTICAL TREATMENTS

Acoustical treatment is the term used for sound or noise control. Acoustical tile and acoustical plaster that are commonly used have sound-absorbing qualities that may be described in specifications; they come in varying types and classes.

- Were acoustic tile containers unopened on delivery? Note: see, Resilient Floor Covering/Tile, above.
- When gypsum board is used as a mounting material for acoustical tile, have the cementing material or fastening devices been carefully inspected to ensure the weight of the tile can be carried? Notes: (1) Gypsum board may be fastened directly to the underside of joists at 16 inches on center; deflection of the floor or roof structure should be limited to 1/240th of the span. (2) If the gypsum board is attached to concrete or steel joists, resilient furring channels at 16 inches or 24 inches will help to dampen sound transmission.
- Have ceiling units been laid out in square patterns and are they symmetrical around the centerline of rooms?
- Are ceiling joints straight, true, flush, and level?
- When acoustical tile was placed on walls, were channels shimmed wherever necessary to obtain true and plumbed alignments?
- Where necessary, has acoustic sealant been used where partitions abut dissimilar floor or ceiling material?

DRYWALL FINISHES

Inspect wallboard upon delivery for conformance with specifications and possible damage during shipment and handling. Gypsum wallboard should be stacked flat and not where it is likely to be damaged.

- Have headers and bridging been installed to provide a firm backing behind all edges?
- Is the contractor installing the wallboard so as to minimize the number of joints? Notes: (1) Use the longest and widest practical panels available to minimize the number of joints. Installation should normally start with the ceiling. (2) Wall installation should start at a corner of the room.
- Is all wallboard held firmly against the framing while nailing? Notes: (1) Care should be taken not to overdrive nails and damage the surface. (2) The bottom should be a minimum of 1/4 inch off the floor.

- Are the size, type of nails/screws and spacing as per specifications? Note: Drywall screws should be driven to penetrate just below the gypsum panel surface without breaking the surface paper. Overdriven fasteners will affect the finished appearance, such as visible depressions over the fastener heads.
- Is wallboards being cut so that they can be set without forcing the boards in place? Note: When cutting or scoring wallboard, the cuts should be made from the finished side of the wallboard; loose panels will create "fastener pops" and joint cracks.
- Are all joints being concealed in a professional manner? Notes: (1) Joint cement should be sanded only as necessary, with care being taken not to scuff the wallboard surface. (2) Improperly installed or missing control joints will not allow for relief of stresses, thereby resulting in cracks.
- Are tradesmen properly maintaining all drywall finishing tools by performing regular maintenance and repair? Note: Replace worn tools and parts as needed.

PAINTING

Many people view painters as being in a "lesser" trade than, say, electricians: not so. Among other things, painters need to know about the chemistry of the paint that they are using and how the particular environment will affect it. They need to understand the dangers associated with the chemicals with which they work, starting with lead. Painters also cover up the mistakes of the other trades.[19] Inspectors too must familiarize themselves with the unique circumstances of the project they are inspecting, surface preparation through the type of coatings to be used.

- Are received materials being properly stored? Notes: (1) All paint materials should be the brand, type, and grade that was previously approved and should be delivered to job in original, unbroken containers with labels and tags intact. (2) Containers should be secured and shelf life should not be exceeded.
- With regards to presurface preparations, are weather conditions satisfactory to do pre-surface preparations, for example, blasting operations?
- Are metal surfaces free of irregularities, for example, weld spatter, slag burrs, sharp edges, and other irregularities? Note: Similarly, all surfaces, for example, wood, plastic, fiberglass, must be carefully prepared or the coating will quickly fail.
- Has all hardware, electrical fixtures, and other accessories been removed prior to painting.
- Are air compressors free of moisture and oil contamination? Note: Compressed air should flow at a steady rate. Pulsations of material normally indicate an inadequate air supply.
- With regards to surface preparations, do the surface preparation methods meet specification? Note: Check especially substrate surface temperature that should comply with specifications until the coating has thoroughly dried/cured.
- Have imperfections and holes been filled or removed?

---

[19] A primary reference used for this section: Bortak, Tom N. *Guide to Protective Coatings: Inspection and Maintenance (Chapter IV and Appendix C)*, United States Department of the Interior, Bureau of Reclamation, Technical Service Center.

- Do all surface profiles meet specifications?
- Are surfaces free of contaminants and do they meet cleanliness specifications? Note: Examine surfaces for such things as rust, mill scale, grease and oil, dirt and dust, soluble salts, water, chalk, and deteriorated coatings.
- With regards to precoating applications, have all coating materials that are on site been approved?
- Are environmental conditions, for example, ambient and surface temperature, humidity, and dew point in accordance with specifications?
- With regards to coating applications, are application methods as specified?
- Are intermediate coats and topcoats being applied within the specified recoat times?
- With regards to post-coating applications, does the dry film thickness meet specification requirements?
- Is the dry film free of pinholes and other imperfections? Note: Ensure that these imperfections are corrected and that the corrections are not obvious.

CONVEYING SYSTEMS

ELEVATORS

See also High-Rise Building Construction, Final Inspections, Elevators: As with, say, well drilling, elevators are usually installed by subcontractors that confine their businesses to elevators, escalators, moving walkways, and the like; they are usually licensed to do this work. Similarly, inspectors who allow elevators into service are normally required to meet the qualifications of the local Department of Building and Safety or similar. The following checklists are provided to help you communicate with the people who install the elevators on your project and to give you an appreciation of elevators, that are sometimes taken for granted. If you are interested (I happen to find then amazing machines) you should study-up on them.

Generally, there are three types of elevators: traction, hydraulic, and machine-room-less (MRL).

Traction elevators are lifted by ropes (commonly referred to by most people as cables) that pass over a wheel attached to an electric motor above the elevator shaft. They are used in mid- and high-rise buildings (say six stories and higher). They have much higher travel speeds than hydraulic elevators. A counter-weight makes the elevators more efficient. Geared traction elevators have a gearbox that is attached to the motor that drives the wheel that moves the ropes. The maximum travel distance for a geared traction elevator is around 250 feet. They can travel at speeds up to 500 feet per minute. Gearless traction elevators have the wheel attached directly to the motor. Gearless traction elevators can travel at speeds up to 2,000 feet per minute and they have a maximum travel distance of around 2,000 feet so they are the only choice for high-rise applications.

A piston pushes hydraulic elevators up; picture a service garage lift. They are used for low-rise applications of 2–8 stories and travel at a maximum speed of 200 feet per minute. The machine room for hydraulic elevators is located at the lowest level adjacent to the elevator shaft. Conventional hydraulic elevators have a sheave that extends below the floor of the elevator pit, which accepts the retracting piston as the elevator descends. Some configurations have a telescoping piston that collapses and requires a shallower hole below the pit. The maximum travel distance for a conventional hydraulic elevator is approximately 60 feet (six stories). Holeless hydraulic elevators have a piston on either side of the cab. In this configuration, the telescoping pistons are fixed at the base of the pit and do not require a sheave or hole below the pit. Telescoping pistons allow up to 50 feet of travel distance. Non-telescoping pistons only allow about 20 feet of travel distance. Roped hydraulic elevators use a combination of ropes and a piston to move the elevator. Maximum travel distance is about 60 feet.

MRL elevators are traction elevators that do not have a dedicated machine room above the elevator shaft. The machine sits in the override space and is accessed from the top of the elevator cab when maintenance or repairs are required. The control boxes are located in a control room that is adjacent to the elevator shaft on the highest landing and within around 150 feet of the machine. MRL elevators have a maximum travel distance of up to 300 feet and can travel at a speed of up to 350 fpm.

- Has the following information been permanently recorded for the owner?
  - The installing subcontractor; obtain the names of the workers.
  - Device number.
  - Rated speed in feet per minute.
  - Rated capacity in pounds.
  - Code data plate information.
  - Crosshead[20] data plate and rope data tags.
  - Numbering: controllers and disconnect switches
  - Any observations deemed relevant.
- Are hoistways of elevators provided with means to prevent smoke and hot gasses from accumulating?
- Are vents properly located? Notes: (1) Vents are usually allowed in the sides of hoistway enclosures below elevator machine room floors or in the roofs of hoistways. Vents should open either directly to the outside or through noncombustible ducts to the outside. (2) Vents are also generally allowed in the walls or roofs of overhead elevator machine rooms through the smoke holes[21] that are in the tops of hoistways; they should be vented to the outside through noncombustible ducts.
- If the hoistways are vented by mechanical means, do the systems provide sufficient capacity to exhaust the code required number of air changes per hour (normally 12)? Notes: (1) Smoke detectors that activate the ventilation system should be placed at the top of hoistways. (2) Ventilation that serves hoistways is usually not permitted to pass through overnight sleeping areas. (3) Manual controls for shutting down the mechanical ventilation systems should be provided in or near the elevator control panel.
- Are machine rooms properly labeled? Note: It is common to require machine room doors to be labeled "*Elevator Machine Room*" with letters ≥2 inches high.
- Does each machine room have a duplex receptacle that is rated at not less than 15A, 120V, and with a ground fault interrupt?

---

[20] Crosshead: The upper member of the car frame.
[21] A smoke hole is an opening for an elevator hoistway venting in the elevator machine room floor at the top of the elevator hoistways.

- Are counterweights located in the hoistways of the elevators that they serve? Note: There may be exceptions to this rule.
- Are hoistways accessible for inspection and maintenance and repair as per code?
- Are hoistways without any windows in the walls and skylights at their tops?
- Is all counterweight material as per code? Note: Counterweights should be of steel, iron, or lead having a minimum melting temperature of 620°F.
- Are emergency stop switches as per code? Note: Emergency stop switches are normally not allowed in passenger elevators but are required in freight elevators.
- Are smoke detectors installed as per code? Note: This requires careful reading of the code.

PLATFORM LIFTS

Platform lifts are permitted under the ADA as a component of an accessible route in new construction in the following applications: performance areas such as stages and speakers' platforms, access to wheelchair spaces and to meet line-of-sight requirements; incidental spaces that are not public spaces; judicial spaces such as witness stands and jury boxes; and existing exterior spaces where exterior constraints make ramps or elevators infeasible. You will frequently find inclined and vertical platform lifts in residences where disabled people live. Check also local building, fire, and elevator codes that may be more restrictive than ADA.

- Do inclined lifts have a minimum capacity of 495 pounds? Note: Vertical lifts require 750–1,400 pounds capacity.
- Do lifts have platform sizes of ≥31½ inches × ≥48 inches?
- Do lifts start/stop smoothly?
- Do lifts slow down before and around corners?
- Do lifts have nonskid platform coatings?
- Do building fire alarm integration systems on inclined platform lifts prevent the lifts from being used during fire emergencies? Note: While sometimes an optional feature, this eliminates the possibility of lifts blocking the stairways.
- Do inclined lifts have the following?
  - Emergency stop buttons on platforms?
  - Emergency manual lowering?
  - Auxiliary power system? Note: This is an optional feature.
- Are the guide rails of inclined platform lifts solidly anchored?

MECHANICAL

Grouped here under this general heading are contractors, generally subcontractors, who install the ventilation, air conditioning, heating, refrigeration systems, and plumbing systems. They often specialize in one trade, for example, plumbing or heating, ventilating and air conditioning HVAC, and in constructing one type of facility, for example, office buildings, hospitals, and so forth.

PLUMBING

The primary functions of the plumbing system are distribution of water and disposal of wastes. Plumbing systems generally include hot-and-cold water piping, drains, wastes, vents, fixtures, insulation, water heaters, and fire protection systems that are located within the building and outside to a specified point of connection with various utilities.

As systems are brought on-line, inspectors should witness all instructions on how to operate and maintain the equipment. These instructions should be complete and understandable; record the names of instructing personnel and those who were instructed.

Water pipe sizes vary between different materials specified despite having the same nominal size. There is a difference between a tube and a pipe; the difference depends on how they are measured, and ultimately how they are used. A pipe is a vessel while a tube is structural. A pipe is measured based on its inside diameter (ID) because it is a vessel, a tube is measured based on its outside diameter (OD) because it is structural. Pipes have a consistent ID regardless of wall thickness. For example, a ½-inch high-pressure pipe may need a 2-inch-thick wall, but the ID will still be ½ inch, even though the OD is 4.5 inches. For all the reasons stated, inspectors should not permit substitution of sizes and types of pipe/tubes, as any substitution may seriously affect operation of the system.

Common wall thicknesses of copper tubing in the United States are Type K, Type L, Type M, and Type DWV.

- Type K is the thickest wall section of the three types of pressure-rated tubing and is commonly used for domestic water service and distribution, fire protection, and so forth.
- Type L is a thinner pipe wall section and is used in residential and commercial water supply and pressure applications.
- Type M is an even thinner pipe wall section and is used in residential and commercial water supply and pressure applications.
- Type DWV is the thinnest wall section and is generally only suitable for unpressurized applications, such as drains, waste, and vent (DWV) lines. For comparison:

NOMINAL SIZE	ACTUAL OD	WALL THICKNESS
K 1"	1⅛"	0.065"
L 1"	1⅛"	0.05"
M 1"	1⅛"	0.035"

- Are all pipes and tubing of the type and grade required by the specifications? Notes: (1) Pipes and tubing should be free from kinks, sharp bends, or dents. (2) Piping and tubing are identifiable by markings placed by the manufacturer on each section.
- Are valves in accordance with the specifications? Notes: (1) The working pressure rating, size, and manufacturer's initials should be cast on the body of each valve. (2) To the extent possible, internal parts of valves should be

inspected to ensure they have not been damaged during shipment, storage, or installation.

- Are all fixtures in compliance with contract specifications and approvals? Notes: (1) Fixtures should be inspected for possible damage such as chips or cracks and should be solidly fastened: fixtures should be clean and ready for immediate use. (2) Valves, such as flush valves, should be adjusted for proper operation.

- Have hot-water lines been installed with the slope as shown on contract drawings or as stated in specifications? Note: Positive circulation of hot water may be provided by the addition of hot water–circulating pumps. These pumps maintain water at fixtures in accordance with the specified temperature setting. Thermostats that are actuated by temperature changes of the domestic-use water can control pumps manually or automatically.

- Have air release valve vents been installed (if required) at high points in the system? Note: Air relief valves are installed to relieve air and prevent air locks.

- When specified, has all cold water piping been installed with a fall towards the shut-off point? Note: This is to allow the entire system to be drained and thus prevent freezing.

- When specified, have air chambers been installed near each valve or faucet? Note: Air chambers are necessary to prevent damage to the system by water hammer; water hammer can actually burst pipes.

- Soil, waste, drain, and vent piping should be installed to provide a gas-tight system. Notes: (1) Soil pipes receive discharge from water closets and similar fixtures while waste pipes receive the flow from fixtures other than water closets. (2) Vent pipes are installed to prevent air locks in lines and to prevent the water seal from being siphoned from traps. (3) Drain pipes receive the flow from waste and soil pipes. This line connects the plumbing system to the building sewer.

- Have soil, waste, and drain lines been installed with a grade of 1/4-inch per foot, where possible, and in no case at a slope <1/8-inch per foot. This fall is required to prevent settlement of the solids.

- Are gutters pitched properly? Note: The gutters that carry rainwater from the roof and other exterior areas should have a grade sufficient to transport sand and dirt particles without having them accumulate in the gutter; this pitch varies depending on rain levels. Check the specifications.

- Are the weep holes that are installed in areas with waterproofed floors above the waterproofing and do they provide drainage for moisture that seeps through the surface of the floor?

- Have vents been installed so that moisture will drain back to the soil or waste pipes by gravity without passing through any traps or pockets in the piping?

- Have all systems been carefully inspected to ensure that there are no cross-connections between the soil, waste, drain, and vent piping and the potable water system?

- Does each fixture have a trap as required? Notes: (1) No fixture should be double-trapped. (2) Trap sizes are determined by the local code. A trap may not be larger than the drainpipe it discharges into. (3) P-traps must be properly vented. Without adequate venting, the trap seal can be removed by backpressure.

- Are all traps level so that the trap seals function properly?
- Are grease traps equipped with devices to control the rate of water flow so that the water flow does not exceed the rated flow of the trap?
- Have shower pans been installed so as to prevent moisture from seeping into materials below the shower floor? Note: Pans that are fabricated in the field should be inspected carefully before the shower floor is placed; look especially for holes in the pan and corners not properly folded.
- Do sump pumps that lift waste to the waste line have the capacity and size that was specified? Note: Bearings should be inspected for grease and floats for soundness (not punctured) and controls to ensure proper operation.
- Have standpipes and hose systems been installed in strict accordance with contract specifications/plans and National Fire Protection Association codes? Notes: (1) Piping should be carefully inspected and tested for leaks before being hidden in walls and ceilings. (2) Caulking and peening[22] joints are not permitted. (3) Hoses should be clean, dry, and unused.
- Do pipes and joints present smooth, rounded corners to the flow of liquids? Note: There should be no mitering of pipes and/or joints for elbows and notching of straight runs of pipe for tees.
- Are union connections accessible? Note: Union connections provide a simple means of disconnecting equipment or fixtures and, therefore, should be accessible and not covered by insulation.
- During installation, are pipe openings capped or plugged to prevent dirt, rodents, and foreign waste from entering the system?
- Does pipe insulation conform to specifications and has it been installed in a neat and smooth manner?
- Are pipe hangers such that they can be adjusted? Note: Check to ensure that temporary supports were removed after lines were completed.
- Have escutcheon plates been installed on bare, exposed pipes passing through floors, finished walls, and finished ceilings? Notes: (1) Escutcheons close the space around the pipe and give rooms a neat and finished appearance. (2) Usually sleeves are required where pipes pass through walls. (3) Specified insulation may be required at firewalls.
- Has all installed piping been tested as per specifications? Notes: (1) Piping should be tested prior to it being painted or hidden by construction. (2) The contractor is usually required to provide all test equipment, unless otherwise specified. (3) Inspectors should witness all tests and record test results in their Project Diary and Daily Log.
- Have domestic water piping systems, both within and outside of the buildings, been sterilized as per specifications?
- Have all electrically controlled items been tested as per specifications? Note: These controls can be very complex and may exceed the knowledge of the construction inspector; don't be afraid to acknowledge this. If necessary, insist

---

[22] Peening is a cold working process in which the surface of the component is deliberately deformed.

that a mechanical or electrical engineer or technician assist in inspecting the electrical hook-ups for electrically controlled valves, thermostats, motors, and so forth.

HEATING: ALL TYPES

This section covers material and equipment for heating systems, including heat generators located within the structure, but does not include central heating plants. Heating system controls are generally highly sensitive, and therefore, all materials, equipment, and installation need to be specified, approved, and inspected prior to and during installation. Generally, approved submittals and details supersede the specifications, as each manufacturer has specifically designed equipment.

- Have all materials and equipment been inspected upon delivery and prior to installation? Materials and equipment must be properly stored and protected. Notes: (1) Reject all damaged material and equipment. (2) The installation of all equipment should comply fully with manufacturer's recommendations; do not accept Gerry-rigged modifications. (3) Unions and flanges should not be insulated. (4) Inspectors should discuss fully the installation schedule including trades coordination. (5) Check nameplate data on all equipment prior to installation. (6) Check accessibility of controls, valves, access panels, and ease of boiler tube removal. (7) Verify that operating and maintenance instructions are available. (8) Check that all spare parts and tools that were specified were delivered and stored properly. Accessory equipment will usually include feedwater controllers, dampers, and pressure and draft gages; flow and pressure recorders; soot blowers, water columns, and boiler blow-downs; and pressure stats.[23]
- Are installation dimensions and clearances for all boilers, furnaces and accessory equipment as per code and specifications? Note: This is necessary for accurate setting.
- Are all joints tight?
- Is the installation free of cracked sections? Note: There should be no gas or air leaks.
- Are refractory brick/mortar, firebox construction, and expansion joints as per specifications?
- Has insulation been properly installed? Note: Ensure that only approved insulation has been used.
- Have all specified safety items been installed? Note: Safety items may include low water cut-off, flame failure, alarms, pop valves, fusible plugs, and backflow prevention if connected to potable water system.
- Do variable air-volume units have shut-off valves to supply and return piping? Note: Ensure that hot water and steam pipes have safety relief valves and rigid discharge pipes.
- Have the combination pressure–temperature gauges been installed in hot water boiler systems, and do the discharge pipes have expansion tubes? Correct as changed.

---

[23] A pressure stat is an electronic device that prevents starting in the absence of water, and the avoidance of water hammer.

- Do steam boilers have equalizing pipes, low-water cut-offs valves, water gauge glasses, and pressure gauges that indicate normal operating ranges?
- Have boilers been tested to maximum system pressure? Note: (1) All accessory equipment and drying of the boiler should undergo testing, even if it will not be put to use immediately. (2) Ensure that manufacturer training is provided to maintenance personnel. (3) Inspectors should witness all tests.
- With regards to oil burners, is the size of the burner as specified?
- Are the locations of electrodes such that ignition of oil is ensured?
- Are pilots accessible?
- Are clearances for removal of burner from furnace adequate?
- Are burners properly adjusted? Notes: (1) Check flue gasses: typical oil burner flue gas contents are nitrogen (N_2), 70–80%; carbon dioxide (CO_2), 12–14%; oxygen (O_2), 2–6%; carbon monoxide (CO), 70–110 ppm; sulfur dioxide (SO_2), 180–250 ppm; hydrocarbons (C_xH_y), <60 ppm; and soot/smoke. (2) Ensure proper combustion air per manufacturer's recommendations and code (generally 40 cubic feet/1000 BTU input).
- With regards to gas burners, are burners clean and properly adjusted and are pilot flames and sensing elements properly positioned? Note: As with oil burners, ensure proper combustion air per manufacturer's recommendations and code, generally (40 cubic feet/1000 BTU input).
- Have gas lines been blown out prior to being connected to burners or regulators?
- Are regulators installed properly?
- Do gasses vent to the outside?
- Are draft fans and drives properly anchored and aligned and do they rotate smoothly, without vibration and overheating?
- Are dampers functioning properly? Note: Ensure that all fans and drives are accessible for maintenance and repair work.
- Has insulation been applied to induce draft fans?
- Are safety control interlocks/air-flow switches functioning properly?
- Are all safety controls and automatic controls installed and do they function properly? Note: Check combustion control operators for support, hookup, and efficiency.
- Are stacks and chimneys properly supported at their bases and are there either guys or supports at intervals, as height requires?
- Are stack plates the correct gauge?
- Are clearances and insulation between stacks and roofs as per code and specifications? Note: Inspect the application and fastening of insulation.
- Are stacks and chimneys the proper elevations, as per code?
- Are feed water treatment and softening of the proper chemical dosages? Note: Witness maintenance instructions and operations.
- If connected to potable water sources, have pressure reducers and backflow preventers as required by code and specifications?
- Do all heating system piping installations allow for expansion and contraction? Notes: (1) Check installation of mechanical expansion joints. (2) Do not remove spacers until joints have been installed. (3) Ensure the proper lubrication and installation of clean filters. (4) Check for kinks, wrinkles, or other defects in pipe bends, the pitch of horizontal runs of pipe,

the position of branch connections, and the use and proper installation of eccentric fittings.

- Have supply and return line shut-off valves that are in heating systems been installed at every heat exchanger, at both sides of pressure reducers, and at valves that are on convectors to pressure vessels and on building supply and returns to central systems.
- Are hangers spaced properly? Note: Openings in walls should not be used as pipe supports.
- Have the insides of pipes been reamed and are threads sharp and of the proper length?
- Are air-release fittings properly located and are they accessible?
- Do all radiator, drip, and boiler return traps that are in heating systems meet their specified capacity requirements? Notes: (1) All condensate and vacuum return pumps should be inspected for nameplate capacity ratings. (2) Drip traps that are at the ends and low points in stream mains, ahead of heating units on long run outs, and at other points where condensate may accumulate should be checked to ensure they are functioning properly.
- Have fittings been installed wherever the gravity flow of vacuum returns is interrupted by changes to a higher elevation? Note: All steam supply and return lines should be cleaned before the systems are allowed to operate.
- Are hot water system lines properly leveled, aligned, and stable?
- Have hot water system lines been inspected to ensure that lubrication, seals, packing, and pressure retention meet specifications?
- Are radiant heating coils accurately placed?
- Are radiant heating coils firmly secured, and have they been tested to ensure water tightness prior to being hidden by construction?
- Have air pockets and restrictions been eliminated? Note: High points should be vented, and the entire system should be able to be drained.
- Have balancing valves and orifices been installed in the return connection of each radiator or heating device? Note: Check the balance and flow distribution.
- Do converters have threaded openings?
- Have safety devices and temperature controls been placed on converters? Notes: (1) Check coils for tightness (2) Ensure that there is adequate clearance to remove coils. (3) Check blow-off safety valves and drainpipes to ensure proper installation.
- Are hot water system circulating pumps properly mounted? Note: Pumps nameplates should be checked to ensure they meet the specifications.
- Are expansion tanks properly sized, and do they conform to code? Note: Check also that the expansion tanks have the correct protective paint coating, insulation, water level gage, drain, and air charging valves.
- Is all heated air from hot air heating systems fully separated from flue gasses? Note: Check return air for free passage back to heater unit.
- Have hot air heating system dampers been properly balanced?
- Have smoke detectors been installed as required?
- Have radiators and convectors been inspected to ensure proper fin tube and baseboard radiation? Note: Inspect also hangers to ensure that expansion can occur without imposing strain on the units.

- Do all HVAC component parts function properly? Note: Check especially: access doors; noise levels; flexible pipe connections and vibration eliminators; and flexible connections to vibrating equipment.
- Do all unit heaters function properly? Check especially: air distribution, noise level, operating controls, and clearances.

VENTILATION SYSTEMS

Ventilation systems provide and remove air to control contaminant levels and humidity and heat and cool the air as necessary. Simplistically, ventilation is usually viewed as being either natural or mechanical ventilation. In natural ventilation, outside air enters the building through opened doors, windows, and other entry points. Mechanical ventilation is usually provided by electrical/mechanical equipment that removes polluted indoor air and brings in fresh air that is then filtered and conditioned and distributed throughout the building. The design of a building's ventilation system offers the opportunity for great savings to owners; ventilation systems may also play an important role in gaining LEED credits in "Earth and Atmosphere," "Indoor Air Quality," and "Innovation in Operations" LEED categories.

- Do fans rotate in the correct direction, and are bearings lubricated, belts tensioned and aligned; and is vibration minimized?
- Do viscous-type filters include washing tanks and spare filters? Note: Check to ensure easy access for filter replacement and that clean filters are installed after final testing.
- Are the service access door of power roof–type ventilators tight? Notes: (1) Make sure that flashing at the curbs is watertight. (2) Ensure that automatic dampers operate properly and that direct discharge of fans is away from air intakes or exhaust outlets. Check local code and specifications for minimum distances from exhaust outlets, intakes, and so forth.
- Do gravity-type ventilators rotate freely?
- Are the blades of back-draft dampers clean and in good condition?
- Are goosenecks and rain hoods properly flashed and braced and do goosenecks face away from the wind?
- Are screens made with the correct mesh size?
- Are ducts made of the correct gauge material? Notes: (1) Check centerline radii at duct bends. (2) Ensure that ducts do not leak and vibrate; specifications may require leakage testing.
- Have duct transitions been constructed properly? Notes: (1) Check to ensure that vanes been installed in right-angle elbows; check specification on this issue. (2) Check transitions for slopes ≤1 in 5 to ensure smoothness, and if a greater slope is necessary, provide splitter vanes. (3) Check for obstructions inside the ductwork. If absolutely necessary, pipes or conduits may have to be run through the ducts in which case they should be encased in streamlined sleeves. (4) If an obstruction is ≥15 percent of the cross-sectional area, then the duct size should be increased to the original cross-section. Check specifications.
- Are access doors at all fire dampers, automatic dampers, coils, filters, heaters, thermostats, and any other item that will require servicing airtight, securely fastened, accessible and can they be fully opened?

- Are drive slide ends cut squarely and are they thoroughly hammered over? Note: Ensure that "S" slips are installed so that the interior piece is in the direction of airflow.
- Are hangers properly spaced and supported?
- Do canvas connections have correctly weighted fabric? Notes: (1) The fabric should be fireproof, have the correct amount slack, and have airtight seams. (2) The canvas should not be painted. (3) The cut edges of all canvas covers should be thoroughly pasted or sewn?
- Have sound-absorption linings been installed properly? Note: Check to ensure that nonflammable cement has been used and that exposed edges cannot break away, crack, peel, flake off, or show evidence of delamination or continued erosion when air is passed through duct sections. Notes: (1)Linings should not have a detectible, objectionable odor. (2) Flame Spread Index = 25; Smoke Development Index = 50. (3) There should be no fungus growth.
- Have butterfly damper rods been marked?
- Do multilouver damper blades open and close properly?
- Do fire dampers close tightly?

AIR CONDITIONING AND REFRIGERATION

This section includes the materials and equipment that are used in air conditioning and refrigeration systems. Construction inspectors can usually inspect airflow systems and evaporative cooling systems, but refrigeration systems are another matter. They can be very complex, in which case inspectors should not hesitate to request the services of mechanical or electrical engineers.

Chilled water has a variety of applications from space cooling to process uses. Industrial chillers cool process fluid or dehumidify air in commercial and industrial facilities. They use either a vapor compression or absorption cycle to cool. Chillers are generally rated at between 1 to 1,000 tons of cooling energy. There are three different types of chillers: (1) air, (2) water, and (3) evaporative condensed. For industrial chillers, there are four subcategories in each of these categories: (1) reciprocating, (2) centrifugal, (3) screw driven, and (4) absorption chillers. The first three types are mechanical chillers that are powered by electric motors, steam, or gas turbines. An absorption chiller is powered by a heat source such as steam and uses no moving parts.

To operate with economy, air conditioning systems must use the refrigerant repeatedly. For this reason, all air conditioners use the same cycle of compression, condensation, expansion, and evaporation in a closed circuit. The same refrigerant is used to move the heat from one area, thereby cooling this area, and to expel this heat in another area.

The refrigerant comes from the evaporator into the compressor as a low-pressure gas. It is compressed and then moves out of the compressor toward the condenser as a high-pressure gas. The gas then flows into the condenser where the gas condenses to a liquid and gives off its heat to the outside air. The liquid then moves to the expansion valve under high pressure. This valve restricts the flow of the fluid and lowers its pressure as it leaves the expansion valve. The low-pressure liquid then moves to the evaporator, where heat from the inside air is absorbed and changes it from a liquid to a gas; evaporators in chillers operate at lower pressures and lower temperatures than do condensers. As a hot low-pressure gas, the refrigerant moves to the compressor where the entire cycle is repeated.

In an ideal cycle, the condenser (from which heat is transferred from the refrigerant to the outside air or water) serves as a two-fold component. Prior to condensation occurring, the high-pressure vapor must be brought to a saturated condition. For this to occur, sufficient heat must be transferred from the refrigerant to lower its temperature to the saturation temperature.[24] Only then can condensation begin. As heat continues to be transferred from the refrigerant vapor to the air or water, depending on the condenser being used, the percentage of the refrigerant that is in the vapor state continues to decrease until the refrigerant has been completely condensed; in an ideal system, this occurs at the outlet of the condenser. But systems are not ideal and therefore some sub cooling usually occurs at the condenser outlet.

At this exit point of the condenser, the refrigerant is in the liquid state and at a high pressure and temperature. To become useful (again) as a heat transfer medium, a reduction in temperature is necessary. This is accomplished by reducing the pressure. A refrigerant's pressure–temperature relationship is an infallible law: when the pressure of a saturated liquid[25] is reduced, it assumes the saturation temperature at the new pressure. Therefore, in order to reduce the temperature, the pressure has to be reduced, and some sort of restriction is required for this to occur. This is what the thermostatic expansion valve does.

The thermostatic expansion valve is an adjustable restriction that causes a reduction in liquid refrigerant pressure, and it modulates to maintain constant superheat at the evaporator outlet. The thermostatic expansion valve is a superheat control, but will not maintain a constant vapor pressure. It provides the restriction necessary to reduce the pressure to some level that will be determined by compressor size, thermostatic expansion valve, load size, load demand, and system conditions. If a constant evaporator temperature is required, it can be achieved by maintaining the pressure corresponding to the required saturation temperature. This is accomplished by adding an evaporator pressure–regulating valve to the system.

In an ideal cycle, there is a pressure drop in the thermostatic expansion valve. Sub-cooling or superheat cannot exist where there is a mixture of liquid and vapor and, therefore, any place in the system where the refrigerant exists in two states, it will be at the saturation temperature for its pressure. Some of the liquid refrigerant must boil to remove the heat necessary to achieve this lower temperature. This is another heat transfer process that yields a lower liquid temperature. The liquid that is sacrificed in the boiling process explains the increase in refrigerant quality. The greater the difference between the liquid temperature and evaporator temperature, the more liquid will have to be boiled in order to achieve the new saturation temperature. This results in an even higher refrigerant quality, that is, a higher percentage of the refrigerant in the vapor state.

The final segment of the refrigerant's journey is as a mixture of saturated liquid and vapor traveling through the evaporator tubing. Warm air is blown across the evaporator, where its heat content is transferred to the boiling refrigerant. This is a latent heat gain to the refrigerant that causes no temperature increase but a change of

---

[24] Saturation temperature: The temperature at which a liquid boils at a fixed pressure, especially under standard atmospheric conditions.
[25] Saturated liquid: A liquid whose temperature and pressureare such that any decrease inpressure without change in temperature causes it to boil.

state. In a perfect system, the last molecule of saturated liquid boils off at the evaporator outlet. The saturated refrigerant then enters the compressor.

In summary: High-pressure gas leaves the compressor and enters the condenser where heat is transferred from the refrigerant to outside air or water. High-pressure liquid leaves the condenser and enters the expansion valve that reduces the liquid's pressure. Low-pressure liquid enters the evaporator where heat is transferred from inside air to the refrigerant. Low-pressure gas leaves the evaporator and enters the compressor that makes it into high-pressure gas that goes into the condenser from which heat is transferred from refrigerant to outside air or water. The cycle continues this way until the refrigerated space temperature is satisfied, and the equipment cycles off.

- Miscellaneous:
  - Is the air flow in air cooled condensers[26] unobstructed and have wind deflectors been installed?
  - Are relief valves in water cooled condensers[27] of adequate size; is spray coverage as per design; and, are there valves for shut off, purge, drain, and liquid level?
  - Have the following items in evaporation cooled condensers[28] been inspected and are they as per specified: spray coverage; float valve operations (noise); water levels; fan rotations and speeds; pump suction strainers; mesh sizes of inlet screens; pans, casings, eliminators, fan corrosion protection, and complete draining; and provision for adjustment of bleed off?
  - Do evaporative coolers have adequate spray coverage, water carry through, the correct water level in sumps, and are float valves relatively noise free?
  - With regards to refrigerant receiver items:
    - Are relief valves sized properly and per specifications?
    - Are valves for shutoffs, drains, purges, and liquid levels sized properly and per specifications?
- Are water coils into refrigeration equipment and piping pitched and vented as per plans and specifications? Note: Inspect also to ensure that the counter-flow of water and air is per plans and specifications.
- Are direct expansion coils installed as per manufacturer's instructions?
- Are all pans of fan-coil units protected against corrosion? Note: Carefully inspect sealed joints.
- Have the following expansion-type water chiller items passed inspection: water drain, vents, correct pass arrangements, and safety devices for freeze protection?
- Have the following flooded style chillers items passed inspection: liquid bleed off at bottom of chillers, high/low pressure protection, automatic alarm devices, temperature controls, and correct-level control adjustments.

---

[26] Air-cooled condensers directly condense exhaust steam from the steam turbine and return condensate to the boiler without water loss.

[27] A steam condenser uses water as the heat-receiving fluid.

[28] In an evaporative cooled condenser, the vapor to be condensed is circulated through a condensing coil that is continually wetted on the outside by a recirculating water system. Air is pulled over the coil, causing a small portion of the recirculating water to evaporate. The evaporation removes heat from the vapor in the coil, causing it to condense. Evaporative cooling differs from typical air-conditioning systems that use compression, condensation, expansion, and evaporation in closed circuit cycles.

- Package type air conditioners:
  - ☐ Are high-pressure cutouts set and functioning?
  - ☐ Are drip pans watertight and draining properly?
  - ☐ Are water regulator valves operational?
  - ☐ Have air filters and strainers been installed as required?
  - ☐ Are thermostats functioning properly?
  - ☐ Are suction and discharge pressures of refrigeration compressors as per manufacturer's instructions?
- Humidifiers:
  - ☐ Have humidifiers for air handling systems been installed correctly? Notes: (1) Familiarize yourself with the manufacturer's installation and operating instructions. (2) Check supports. (3) Be sure that the steam supply corresponds to the pressure stamped on the humidifier. Ream pipe and blow out at full steam before connecting to humidifier. (4) Ensure that specifications account for expansion. (5) Ensure condensate is disposed off as per plans and specifications.
  - ☐ Do dry steam grid-type humidifiers comply with all applicable codes and specifications? Notes: A dry steam grid-type humidifier should normally be provided with the following: (1) Steam jacketed separating chamber. (2) Stainless steel dispersion tube with internal silencing screen, sized by the manufacturer to fit the duct at the required humidifier capacity. (3) Spring electric-actuated fully modulating control valve. The control valve is at the normally closed position on loss of electrical power to the actuator. (4) Float and thermostatic type (inverted bucket steam trap if above 15 psi) drain trap, pipeline strainer, and escutcheon plate to completely seal the duct opening. (5) Duct-mounted airflow switch option. (6) Condensate return thermostatic switch option set at a temperature of 205°F. (7) Steam strainer.
  - ☐ Are all dry steam grid-type humidifiers installed so as to comply with all applicable codes and specifications? Notes: (1) All installations should comply with the manufacturer's published recommendations. (2) Galvanized steel rods should be used to support distribution manifolds and mount in air system plenums.
  - ☐ Are the dry steam grid–type humidifiers that are installed in central air handling units on the downstream side of coils or final filter banks? Notes: (1) If the humidifier must be located within the air handling unit casing, the humidifier should first be shipped to the air handling unit manufacturer to ensure that the humidifier is properly sealed when the low air leakage test is being performed on air handling unit by the manufacturer. (2) The humidifier should have sufficient distribution manifolds to limit the steam particle vapor trail to less than 3-feet at scheduled capacities. The number of distribution manifolds and duct size dimensions determine whether vertical or horizontal mounting configuration will be used. (3) Humidifiers that are installed in ducts should have sufficient clearances for maintenance. The duct section where the humidifier is located should be constructed of stainless steel material. The bottom of the stainless steel duct section should be pitched to drain condensed steam pipe; a stainless

steel National Pipe Thread drain connection welded to the bottom of the duct to the nearest sink or floor drain.

- Compressors:
  - □ Do all high- and low-pressure static switches, and oil pressure failure switches operate properly?
  - □ Are oil levels and oil viscosities per manufacturer's instructions?
  - □ Are the amounts and types of refrigerant charges per manufacturer's instructions?
  - □ Do compressors hold pressure upon pump down?
  - □ Do installations meet isolator deflection and compressor vibration requirements?
  - □ Do suction strainer screen meshes meet equipment requirements?
  - □ Are all tensions, alignments, and rotations of belts per manufacturer's instructions?
  - □ Are compressor motor amperages while under maximum load as specified?
  - □ Are refrigerants such that floodbacks and oil foaming do not occur?
  - □ Do cylinder heads operate without overheating?
  - □ Are there automatic oil heaters in crankcases?
  - □ Are there loops in refrigerant piping and condensers for oil traps?
  - □ Are compressors, drives, and gear boxes in proper alignment?
  - □ Do suction dampers or inlet vanes operate satisfactorily?
  - □ Do safety control circuits operate satisfactorily?
  - □ Do surge compressors operate satisfactorily?
  - □ Do float valves operate satisfactorily?
  - □ Do oil pumps and oil coolers operate satisfactorily?
  - □ Are noise and vibration at acceptable levels?
  - □ Are air compressor tank drains conveniently located?
- Absorption refrigeration machines:
  - □ Were all parts kept clean during installation?
  - □ Do liquid level controls operate satisfactorily?
  - □ Do purge pumps operate satisfactorily?
- Refrigeration piping:
  - □ Is piping sized correctly and of the specified materials?
  - □ Has refrigeration piping been inspected and deemed acceptable:
  - □ Has the refrigeration piping been cleaned? Notes: (1) Cleaning should be done before erection. (2) Joints should be cleaned before soldering, in accordance with applicable procedures for refrigerant piping.
  - □ Are pipe hangers isolated from the pipes by a strip of lead or rubber?
  - □ Are flexible metal hoses that are installed for vibration absorption at right angles to the motion in the pipes?
  - □ Were systems dehydrated as recommended by the manufacturer?
  - □ Were systems charged with the correct type and amount of refrigerant?
  - □ Are all systems tight throughout?
  - □ Have the specified number of oil traps been installed?
- Other refrigeration specialties:
  - □ Are expansion valves for superheat settings and for bulb and equalizer positions in accordance with the manufacturer's recommendations?

□ Are solenoid valves installed such that the arrows on the valve bodies are in the direction of flows through the valves? Note: If a valve is backwards, the flow will not be stopped when the valve is electrically de-energized.

□ Are sight glasses clean, accessible, and easy to read, and can cap gaskets be replaced relatively easily?

□ Do evaporator pressure regulators function properly?

□ Do holdback valves for operations during start-ups function properly?

□ Are float valves or switches mounted level and at a height that will ensure correct liquid levels in the evaporators?

□ Are refrigerant drier canisters air tight prior to opening? Note: Piping for replaceable-type dryers should be installed so as to facilitate replacement.

□ Are pipe connections for liquid suction heat exchangers as specified?

- Unit cooler items:

  □ Are corrosion-protected pans and casings made of the specified materials and have they been installed properly?

  □ Do water defrost units receive complete spray coverage, but with no carryover?

  □ Are electric defrost units for cycle timing in accordance with the specifications and job conditions?

  □ Are hot gas defrosts for suction pressures and refrigerant charges in accordance with manufacturer's recommendations?

  □ Is drainage complete during defrost cycles? Note: Ensure that cycle timing is set as specified?

- Natural draft cooling towers:

  □ Are guy wires, corrosion-protected bolts, watertight basins, cracked or split air louvers, spray coverage, and float valve per specifications?

  □ Is water maintained at a constant level?

  □ Are air intakes, fan rotations and speeds, belt tensions, and weather protection of motors as specified and per manufacturer's instructions?

  □ Does water flow through outlets without forming vortices that draw air in with the water?

  □ Do temperature controls and drainage devices comply with specifications?

- Pumps:

  □ Are pump settings as per specifications?

  □ Is all packing or are mechanical seals as per manufacturer's instructions? Notes: (1) Check couplings to ensure alignment is correct and vibration minimized. (2) Check motors for rotation and weatherproofing.

- Controls:

  □ Have control instructions, including sequence of operations, and complete schematic control drawings been furnished? Notes: (1) The instructions should be used during the final acceptance test. (2) Check each function of the controls. (3) Ensure that all control wires are color-coded.

  □ Have damper motors been inspected with fans running?

  □ Do all control valves close tightly?

  □ Have cycle times been checked with all controls operating?

  □ Are all duct-mounted controls airtight and accessible?

  □ Are pneumatic systems airtight and restriction free?

  □   Are electronic system cables and shielded cables properly grounded?
  □   Are amplifiers sited away from magnetic fields, such as large transformers?
  □   Are graphic panels free of damages and dirt between plastic and back plate? Note: Check also for missing control wires and accessibility to all controls.
  □   Has all electrical equipment been inspected for interlocking?
- Insulation:
  □   Are metal bands tight and spaced properly?
  □   Are the cut edges of all canvas covers thoroughly pasted or sewn?
  □   Does strapping have corner protectors?
  □   Is pipe and ductwork insulation tightly sealed?

TESTING AND TRAINING (HVAC SYSTEMS)

Almost all purchases of major pieces of equipment include the training of the operations and maintenance staff; this short checklist is for such training. Nowadays, however, owners are more and more turning to commissioning. Commissioning is usually accomplished by a third-party firm and is far more extensive than a usual acceptance inspection. You might compare commissioning of a new building to the commissioning of a ship, that is, a complete shakedown of all systems over an extended period. When a building is initially commissioned, it undergoes an intensive quality assurance process that begins during design and continues through construction, occupancy, and operations. Commissioning ensures that the new building operates initially as the owner intended and that building staff are prepared to operate and maintain its systems and equipment. While on the job, an inspector's role is to cooperate fully with the engineers who are doing the commissioning and to learn.

With or without commissioning, inspectors should ensure that proper training is provided to operations and maintenance O & M personnel as per the requirements of the contract. Mechanical inspectors should witness all tests and training; record the dates, times, and names of all the attendees; record nameplate information of the equipment in question; and make this all part of the permanent record.

Some specifications may state that any component, device, or unit that bears a U.L. label should be considered in compliance with the local code and, therefore, needs no further testing.

GAS DISTRIBUTION

This section includes the conveyance of natural or artificial gas to outlets, equipment, and appliances within a building from a specified point of connection with the service line. The gas is usually used for heating of facilities and heating water for domestic use. It is common that no gas distribution piping containing gas at a pressure in excess of ½ psig[29] be allowed to run in a building except for commercial use, industrial use, or other large-volume use.

---

[29] Gas pressure in building pipes is often measured in inches of water while the pressure in the mains or service line is usually measured in pounds per square inch gauge (psig). There are 0.0361272918274 psig in 1-inch water [4°C]. Here it is assumed that the conversion is between psig (**pound/square inch [gauge]**) and **inch water [4°C]**.

Many gases are odorless and colorless and burn with a luminous flame; therefore, for safety odor is added to enable its detection.

All materials that are used in gas service and meter piping systems should be in accordance with the requirements of the local utility company; pipe and fittings should be checked to ensure that they have the proper markings and are of the specified material.[30] It is common that the permissible stress for gas piping is limited to not more than 20% of the yield strength of the piping, including primary and secondary loads. Many contracts list the Federal Specifications applicable to materials and the National Board of Fire Underwriters Standard (NBFU) No. 54. Plastic pipe is usually limited to polyolefins that conform to ASTM D2513-1976, Thermoplastic Gas Pressure Pipe, Tubing, and Fittings. Contract specifications may also require coding of gas piping in accordance with ANSI Scheme for Identification of Piping Systems (ANSI A13.1-1956). Inspectors should become familiar with these specifications and codes.

Usually, contractors who install gas installations must be specially licensed. Then the city/county will inspect the installation; a pressure test would usually be part of this inspection. Then, it is common that the gas company or utility will make its own inspections of all facilities before connecting the meter and admitting gas to the system. The gas company or utility will often make a second inspection of all facilities after gas is admitted to the system to assure the proper functioning of the systems and all appliances.

- Has the contractor performed all tests as set forth in the specifications and as recommended by the manufacturer? Note: All tests should be witnessed and documented and made part of the permanent records.
- Have all pipes and fittings been stored so as to ensure cleanliness?
- Has all equipment been inspected for compliance with contract drawings and specifications? Notes: (1) Labels on the equipment should normally show that the American Gas Association has approved it. (2) Any equipment that has been damaged should be replaced.
- Are fittings made of malleable iron? Note: Malleable iron is cast iron that has undergone an extended annealing process to increase ductility.
- Do installations comply with contract plans and specifications, NBFU No. 54 and approved shop drawings?
- Have drips for the removal of condensate been installed as required? Drips should be installed in accordance with NBFU No. 54.
- Have pipe cuts been reamed and all burrs removed before threading? Notes: (1) Clean, sharp cut, undamaged threads of sufficient length produce tight joints. (2) Gas distribution piping operating at a pressure of over ½ psig to 3 psig and ≥4 inches should be welded. All gas distribution piping operating at pressure over 3 psig should be welded.
- Is the thread joint compound that was used an oil and graphite, or graphite compound? Notes: (1) Properly cut threads require lubrication when being tightened; sealing compound should not be allowed. (2) The application of oil and graphite compound should be to the male thread only. This prevents the compound from being forced into the pipeline.

---

[30] Galvanized pipe is usually not specified for gas systems. However, galvanizing may seal minute pinholes that would otherwise leak.

- Are pipes supported by approved adjustable hangers that are properly spaced and aligned?
- Does any manifold that has been installed in gas burning units that are adjacent to each other provide a sufficient volume of gas at the same pressure for both units? Note: The gas supply should be connected to the manifold at both ends.
- Have gas pressure regulators been checked against the approved shop drawing and for operation as required by specifications? Notes: (1) Vents, where required, should be run to the outside of buildings and comply with all code requirements for vents, for example, distances to windows and doors. (2) Verify that regulators are adjusted to deliver gas at the required pressure.
- Do all valves operate freely and are all lubricated plug-cocks lubricated with the correct type of lubricant? Note: Usually, all valves on the gas line should be of the plug-cock type.
- Are all water heaters in compliance with contract plans, specifications and approved shop drawings? Notes: (1) Looking at each heater individually, the capacity recovery rate and temperature rise should be the same, or greater than specified. (2) Back-draft diverters should be installed to prevent pilot lights from being extinguished. (3) Equipment and controls should carry the American Gas Association approval label.
- Do the types and capacities of relief valves conform to specifications? Note: Shutoff or check valves should not be installed between the relief valves.
- Are flues smoke tight and clear of combustible construction? Note: Refer to NBFU No. 54[31] for requirements for clearance.
- Are orifice jets properly sized for the gas used, altitude, etc.?

ELECTRICAL

The standard reference code for electrical work is the National Fire Protection Association's NFPA National Electrical Code that may have been modified by local authorities and sometimes by the specifications.

INTERIOR

All electrical work should be inspected before it is covered. Poor workmanship and non-approved materials will usually become apparent only after the project has received its final approval.

- Has the contractor submitted shop drawings and/or catalog cuts as required? Note: Usually, at least one copy of the approved submittals should be available on the job site for use in inspecting prior to installation.
- Has the contractor complied with all comments that were written on the shop drawings?
- Are all electrical components being stored where dust and moisture will not damage them? Notes: (1) Some electrical components such as control relays,

---

[31] NBFU 54: Regulations of the National Board of Fire Underwriters for the Installation Maintenance and Use of Piping and Fittings for City Gas as Recommended by the National Fire Protection Association.

and certain switching devices are delicate and require careful handling. (2) All electrical materials should be protected from moisture.

- Are lighting fixtures being stored so as to eliminate damage to finishes and reflective surfaces?
- Do the grounded conductors of interior wiring systems have a color identification that is used continuously throughout the system? Notes: (1) The colors, white or natural gray, are usually used for identifying the neutral or grounded conductor on insulated conductors, No. 6 gage and smaller. On conductors larger than No. 6, the neutral should be distinctively marked at the terminals. (2) Bare aluminum or copper-clad aluminum grounding conductors should not be allowed when those are in direct contact with masonry or the earth or where they are subject to corrosive conditions. When used outside, aluminum or copper-clad aluminum grounding conductors should not be terminated ≤18 inches of the earth. (3) Grounding electrode conductors should be installed in one continuous length without a splice or joint, unless spliced only by irreversible compression-type connectors that are specifically listed for the purpose or by the exothermic welding process.[32]
- Are the specification clear as to which conduit wiring method should be used on the project? Notes: (1) The NEC has approved numerous wiring methods, but in each contract the wiring method should be clearly defined. In most cases, polyvinyl chloride, rigid metal conduit or electrical metallic tubing is required. Rigid conduit provides protection to the conductors and acts as an effective grounding conductor for equipment. Rigid conduit may be used under most conditions. (2) Electric metallic tubing (EMT) may be approved for use under the same conditions as rigid metal conduit, except that it may not be used in exterior locations in any size larger than 2-inch nominal internal diameter and may not contain any conductor operating at a voltage higher than 600 V. (3) Conductors of circuits rated over 600 V, nominal, should not occupy the same equipment wiring enclosure, cable, or raceway with conductors of circuits rated 600 V, nominal, or less, unless specifically permitted by the code. (4) To avoid damage to insulation or stretching wires, a run of conduit or EMT between any two outlets, between any two fittings, or between an outlet and a fitting, should not contain more than the equivalent of four 90° bends or quarter bends (360° total), including those bends located immediately at the outlet or fitting.
- Has all damaged conduit and EMT been replaced?
- Is all conduit and EMT well supported?
- Is all conduit and EMT secured with locknuts at each outlet and junction box? Note: This effectively makes a continuous grounded conduit and equipment system.
- Is each wire or cable conductor marked or stamped at intervals on the insulation giving the size and type of wire? Note: This should be checked against contract specifications and/or drawing requirements. Check especially that the number

---

[32] Exothermic welds have higher mechanical strength than other forms of weld and they have excellent corrosion resistance qualities. They are highly stable when subject to repeat short-circuit pulses and do not increase electrical resistance over the lifetime of an installation.

of wires in any conduit is in accordance with tables in the National Electric Code.

- Have raceways that are subject to thermal expansion and contraction been provided with expansion fittings? Note: Where portions of a cable raceway or sleeve may be subjected to different temperatures and where condensation may be a problem the raceway or sleeve should be filled with an approved material to prevent circulation of warm air to a colder section of the raceway or sleeve.
- Have the following tests been conducted? Note: The specifications should be checked for additional tests that may be required:
  □ Fixtures should be checked for proper operation;
  □ Motors should be checked for proper rotation and overheating;
  □ Each convenience outlet should be checked with a meter;
  □ Remote control circuits should be checked for function and operation. Note: More complete and complicated tests and inspections may have to be accomplished by an electrical engineer.
- Have electrical installations that are in hollow spaces, vertical shafts, and ventilation or air-handling ducts been installed so that they will not increase the possible spread of fire and products of combustion? Note: Openings around electrical penetrations through fire-resistant walls, partitions, floors, or ceilings should be fire-stopped using approved methods and materials.
- Have metal wire ways been installed so that their manufacturer's name or trademark are visible?

EXTERIOR

Often a public utility company or the equivalent will do much of this work. If this is the case, an inspector should make sure that all materials that are supplied to the utility company meet power requirements for the completed project. Usually the specifications specify the responsibilities of the contractor and utility company. From an inspection perspective, all contractor-installed exterior work should be inspected as thoroughly as the interior work. Inspectors should not hesitate to obtain the services of an electrical engineer when electrical work is proceeding at a rapid pace or is complex. If signage is being installed, be aware of the code sections that pertain specifically to signs; also, there may be special permits required for the installation of exterior signs.

CONSTRUCTION SAFETY

The contractor and subcontractor(s) must comply with all OSHA regulations applicable to the construction project. No contractor should require employees to work under conditions that are unsanitary, hazardous, or dangerous to their health or safety. On larger jobs, it is usual for each contractor/subcontractor to initiate and maintain an effective safety program. Their programs should include systematic policies, procedures, and practices to protect workers from, and allow them to recognize, job-related safety and health hazards. The programs should include provisions for the systematic identification, evaluation, and prevention/control of general work site hazards, specific job hazards, and potential hazards that may arise from foreseeable conditions. On large projects, it is usual for an owner to require a competent person to conduct frequent and regular inspections and sometimes have such a person on site full time. Each employee should be instructed on how to recognize and avoid unsafe conditions and the regulations applicable to their work environment.

While the contractor has ultimate responsibility for safety during construction until final completion and acceptance of the project by an owner, construction inspectors are ideally situated to identify hazards and should constantly be on the alert for them. Inspectors should review and understand all contract safety requirements before work begins. There should be an understanding with both the owner and contractor that if at any time during construction the inspector recognizes a condition, method, or practice that constitutes an imminent danger,[33] the inspector can stop the work and immediately notify the on-site supervisor and safety personnel.

PRECONSTRUCTION SAFETY

As with all other checklists in this book, this checklist is not all inclusive, but it serves as a tool to assess a project's safety program. As the project progresses, tasks change, and new risks and hazards are introduced, inspectors should address these additional safeguards and corrective measures.

- Do the contractor and his/her key on-site managers/supervisors understand their responsibilities for compliance with contractual safety requirements?
- Has the contractor prepared a Safety Plan that analyzes anticipated hazards of the project and indicates how the hazards will be controlled?[34] Notes: (1) Contractors should identify specific procedures for controlling hazardous operations such as crane use, scaffolding, excavation/trenching, and so forth, (2) Contractors/subcontractors should especially address the four leading hazards of construction work: falls, struck by, caught in/between, and electrical.
- Has a "Designated Competent Person," that is, a person who is responsible for the project's safety and health performance, been appointed? Note: This person should be knowledgeable in applicable OSHA regulations and capable of identifying existing /predictable hazards.
- Have contractor and subcontractor employees been instructed as to what to do in the event of an emergency such as injury, fire, accident, hazardous material

---

[33] Imminent danger: As used here, an immediate threat of harm, which varies depending on the context in which it is used. For example, the existence of any condition or practice on a construction site that could reasonably be expected to cause death or serious physical harm to any worker if construction operations continued in the affected area or if workers were to enter the affected area before the condition or practice was eliminated would be considered an imminent danger.

[34] Such a plan may include accountability; audits/inspections; cell phone usage; communication; competent person; concrete/masonry; confined spaces; cranes and hoists; demolition; electrical safety; environmental and occupational health; equipment safety; excavation and trenching; fall protection; fire prevention and protection; hazard communication; incident management and prevention–emergency action plan; job hazard analysis; ladders; material handling; moisture control and mold prevention; personal protective equipment (PPE); pest management; public protection; recordkeeping and reporting; safety meetings; scaffolding; sign, signals, and barricades; steel erection; substance abuse policy; temporary elevator usage; tool safety; training; visitor policy; welding/cutting. (List extracted from The University of Michigan, *Construction Safety Requirements*, January 2010 (revised 1.3.12)).

spill, and so forth? Note: These instructions should include who to contact, a means of communication with contact numbers, how to contact first-aid/ medically trained personnel with appropriate medical supplies, and other site-specific information.

- Have contractor/subcontractor employees been instructed as to confined space entry, scaffolding, fall protection, hazard communication, lockout/tag out procedures, respiratory protection and First Aid.
- Have evacuation routes, signals, and procedures been provided to all on-site personnel.
- Have safety and health information been disseminated to all site personnel?
- If a contractor/subcontractor's trailer has been brought on site, has it been sited and hooked-up properly? Notes: (1) Hook-ups generally include telephone, electrical, and propane. (2) Adequate space should be provided for vehicle parking. (3) If the trailer is to serve as a first-aid station it should be so marked. (4) Other warning signs, for example, "Hard Hat Area," "Construction Site— CLOSED," "Area Closed to Public," and so forth, should also be posted. (5) Instructional signs, for example, designated parking area, and so forth, and flagmen or barriers may also be required?
- Are construction areas clearly delineated and marked?
- Is personal protective gear (e.g., eye/hand/hearing/respiratory) available to all who need it?
- Are powered tools and equipment in good working order and stored properly? Notes: (1) Belts, pulleys, gears, machine parts, and so forth should be properly guarded. (2) Owner's manuals for equipment should be available to all who need them. (3) Slip hazards should be addressed as required. (4) Emergency equipment should be available and in good condition. (5) Proper signage of hazards and work procedures should be posted. (6) There should be inventory and associated Material Safety Data Sheets available to all. (7) Chemicals, particularly flammable liquids, should be stored properly and clearly labeled. (8) Chemical waste disposal procedures and supplies in place and known to all. (9) Compressed gases should be seismically anchored, well labeled, and in good condition. (10) Electrical equipment, particularly cords, should be in good condition. (11) Lockout/ tagout signage and kits should be available. (12) Fall protection measures should be in place. (13) Ladders and steps should be in good condition. (14) Where required, seismic bracing should be installed for cabinets, bookcases, equipment, and so forth. (15) There should be clear exit paths and aisle ways. (16) A high standard of housekeeping practices should be maintained, always (see below).

HOUSEKEEPING STANDARDS

- Are work areas free of form/scrap lumber with protruding nails and all other debris?
- Are all combustible scrap and debris being removed at regular/frequent intervals?
- Are materials being stored with regard to their fire characteristics?

- Are enclosed chutes being used whenever materials are dropped more than 20-feet to an exterior point of a building?
- Are aisles, exits, and passageways being kept clear and in good repair?
- Are contractors/subcontractors providing potable water, adequate toilet facilities, and adequate washing facilities for employees who are engaged in operations involving harmful substances?
- Are intoxicating beverages, controlled substances, and firearms banned from the site under all circumstances?
- Are all activities in and around water bodies in compliance with regulations for worker safety, including personal flotation devices or buoyant work vests, lifelines, ring buoys, and guardrails while working over or adjacent to water hazards, and from barges, boats, and/or launches?
- Construction site:
  - Do contractors/subcontractors require workers to wear appropriate personal protective equipment (PPE)? Notes: (1) Contractors/subcontractors are responsible for ensuring that workers who use protective equipment are trained and medically qualified to do so. (2) Included under the term protective equipment are eye and face protection (safety goggles, safety glasses, welding goggles, face shields, etc.), protective footwear (safety shoes, boots), and protective gloves (leather, rubber, insulated). Protective masks (dust, half-face, full-face, etc.), hearing protection (ear plugs, muffs, etc.), Coast Guard-approved personal floatation devices PFD/buoyant vests, and fall protection equipment.
  - Are hand tools, power tools, and jacks maintained in safe operating condition and used only for the purpose for which they were designed?
  - Are electric power-operated tools either approved double insulated or properly grounded? Notes: (1) Ground Fault Circuit Interrupters or assured equipment-grounding conductor programs should be in place. (2) Electrical cords/insulation should be in good condition. (3) All ground plugs should be on electrical cords, if so designed by the manufacturer. (4) Workspaces, walkways, and similar locations should be clear of electrical cords. (5) All electrical extension cords should be of the three-wire type. (6) Contractors/subcontractors should have aggressive Lock-out/Tag-out programs wherever required.
  - Do cranes and derricks comply with applicable standards? Notes: (1) Rollover protection should be on all equipment, as required. (2) Back-up alarms or observer signals should be on all equipment, as required. (3) Crawler-type vehicles should not be allowed on improved roads.
  - Do all ladders and their use conform to safety standards?
  - Do aerial lifts and their use conform to safety standards?
  - Has an assessment been made to determine if fall protection is required? Note: (1) Check especially scaffolding, stairs, ladders, and so forth. (2) Toe boards should be installed along the edges of the overhead walking/working surface. (3) Safety nets should be installed to comply with standards. (4) Railing must conform to standards.
  - Are contractors/subcontractors aware of and do they understand the need to incorporate applicable safety precautions for concrete and masonry construction, welding, and cutting operations (e.g., compressed gas

cylinders are protected from vehicle traffic, valve caps secured), hazardous materials/waste handling and storage (e.g., paints, solvents, fuels), and use and storage of flammable/combustible liquids?

- Are proper procedures in place for managing the use, storage, and disposal of hazardous materials?
- Are wastes and containers being properly disposed off per environmental regulations?
- Are the hazards of product use and the need for protective measures (e.g., PPE, etc.) being told to workers?
- Does work with/on asbestos conform to applicable standards?
- Is exposure to toxic gases, vapors, fumes, dusts, and mists being managed correctly?
- Are lead, lasers, and any other hazardous components managed correctly?
- Are subcontract employees being held to the same standards as regards safety as contractor employees?

FINAL INSPECTIONS

As with the final inspections for light construction, the final inspections for high-rise construction are somewhat anti-climatic. If the contactors, subcontractors, and inspectors have been doing their jobs during construction, the final inspection is a joyous event with the owner, CM, contractor, architects, engineers, government inspectors, and indeed all the participants present. Nevertheless, a final inspection is necessary. The players will want to determine that the work that was accomplished and the materials and equipment comply with the specifications and codes. Final inspections are usually conducted in two phases: (1) a pre-final inspection and (2) a final inspection.

It should be clear who has the authority to give final approval and accept a completed project. For government projects, this is usually the Contracting Office. Private owners must make it clear who will sign off on the project. Whoever it is, this person is authorized to reject or disapprove shoddy and defective materials, equipment, and workmanship, and require the contractor to correct or replace what is being rejected. Regardless of who will be signing-off, you can be sure that this person will rely heavily on the construction inspectors, other field inspectors, and specialty inspectors, for example, elevator inspectors.

The pre-final inspection serves a dual purpose: (1) to ensure that the project is ready for a final inspection and (2) preparation of a list of deficiencies, aka a "Punch List." Items that you may wish to consider for your pre-final inspection checklist are provided.[35] They include commonly found deficiencies, but the list is in no way all-inclusive. Besides what is provided, you should review your notes, daily log, plans, specifications, codes, and so forth. Add to your list those items that you had planned to get back to, but didn't; after investing a sizeable part of your life on the project, you owe it to the person who has been paying you and to yourself to carefully plan this final inspection. You do all this so that the final inspection is pretty much a formality, a joyous event.

---

[35] Main references include: NYC Building Code, High Rise Inspection Guide, City of Huston, TX, Salt Lake City Fire Prevention Bureau, High-Rise Inspection Checklist.

ARCHITECTURAL

- Are all Fire Department connections approved, identified, and unobstructed?
- Is a key identification system in place? Note: This is important and sometimes complex in large facilities.
- Are building addresses plainly visible and legible from streets?
- Do all exterior exits have approved discharge landings?
- Are fueled equipment stored properly?
- Does caulking comply with specs?
- Are doors, inclusive of hardware, finish, closers and stops, as per the door schedule? Note: Check to ensure they open and close properly.[36]
- Are windows, including hardware and finish as per window schedule?
- Does paint cover painted areas?
- Are tiles clean and sealed? Note: repair/replace as required.
- Is masonry as per contract? Note: Check especially mortar joints; point as necessary.
- Is drywall free of cracks, holes, and so forth? Note: Repair as required.
- Is metal trim as per contract? Note: Check especially for sharp edges, finish, and so forth.
- Are floors finished properly?
- Are all millwork, cabinet doors, counter tops, and so forth, finished properly?
- Are fascia boards free of roofing materials such as tar?
- Where required, is there an approved assembly permit?
- Are all Fire Department connections accessible, properly identified, and FD inspected and approved?
- Roof, penthouse level:
  - Are all flashing, scuppers, parapet wall coping, and so forth, as designed?
  - Are standpipe manifold valves at roof level; at upper standpipe terminus?
  - Are standpipe pressure gauges at roof manifolds or standpipe termini?
  - Do standpipe pressure gauges have double-drain assemblies?
  - Do gravity tank(s) provide water to standpipes?
  - If the building emergency generators are located on roofs, does fuel storage comply with building code?
  - If there are heliports on roofs, are there provisions for spill containment? Notes: (1) Ensure that fire protection for the heliports comply with local code. (2) If possible, observe a helicopter land and takeoff. Look for possible flying debris. Ensure that the roof is secure and that no ballast can be blown off the roof.
  - Has the local Fire Department approved the roof accesses?

---

[36] Door swing terminology may differ depending on supplier or region. However, usually the direction the door is determined while standing on the outside. If you stand on the outside of the door and you push it away from yourself to open, it is in-swing. If you stand on the outside of the door and you pull it towards you to open, it is out-swing. To determine hand, stand on the outside side of the door. While facing the door, if the hinges are on the right side of the door, the door is "Right handed"; if the hinge is on the left, it is "Left handed." If the door swings toward you, it is "Reverse swing"; or if the door swings away from you, it is "Normal swing." If you stand on the outside of the door and you push it away from yourself to open, it is in-swing.

- Do elevator penthouses/HVAC equipment rooms comply with building codes?
- Do fire extinguishers and fire hoses comply with building codes?
- Do exit signs comply with local codes?
- Are roof tops with exposed electrical wiring per code?
- If stairwell pressurization or smoke removal systems have been installed, do they operate as per design?
- Are air handlers and electrical and mechanical equipment rooms free of combustible materials?

- Exit stairwells:
  - Are 2-hour fire ratings maintained throughout all enclosed exit stairway systems?
  - Has emergency lighting been installed in exit stairwells? Notes: (1) Ensure that directional signs are installed and that stairwells are free of obstructions. (2) Inspect approved electronic egress or exit devices on doors. (3) Inspect all required signage at the stairwell terminus and roof access.
  - Do exit stairways discharge to the outside? Note: Stairways should prevent people from running down into basements/cellars.
  - Are approved exit door devices on exit-level discharge doors?
  - Are approved discharge areas and routes equipped with emergency lighting throughout?
  - Are standpipe hose valve thread protective caps in place?
  - Pressure-reducing valves/devices provided on stairway?
  - Are floor penetrations sealed with 2-hour fire stop materials?

- Exit corridors:
  - Do assigned fire ratings match legal occupancy?
  - Are exit signs in place, as per code?
  - Are corridors free of obstructions?
  - Are elevator lobby evacuation signs and graphics per code?
  - Have electronic locks on corridor doors and elevator lobby doors been permitted?
  - Do elevator lobbies have adequate and approved exiting?
  - Has approved emergency exit way lighting been installed throughout? Note: Emergency exit ways should be connected to the emergency generator system.
  - Are fire hose/extinguisher cabinets unobstructed?
  - Do fire-rated doors in corridors, stairwells, and so forth, have automatic door closures.
  - Are all electrical and mechanical rooms and janitorial closets properly identified?
  - Are all control boxes, throw switch boxes, and breaker switch panels identified and labeled?
  - Do lights have approved bulb protection?

- Tenant areas:
  - Are the criteria listed under exit stairwells and exit corridors met?
  - Is adequate approved exiting provided? Note: Exiting should not be through storerooms, kitchens, mechanical rooms, or other hazardous areas.

☐ Are approved door locking devices installed?

☐ Are clothes dryers properly vented?

☐ Are ceiling tiles in place in a manner that prevents fire/smoke from spreading and to increase the efficiency of smoke detectors and fire suppression systems?

☐ Are approved trash, storage and recycling bins available as per code? Note: Rubbish/linen chutes may require sprinklers.

☐ Do stairwell doors have free egress from tenant spaces into stairwells? Note: Doors to apartments and private offices should be locked from the stair side.

- Parking facilities:

☐ Are spaces properly lined, as per the approved plans?

☐ Is lighting adequate to ensure safety?

☐ Are exit signs visible and in working order?

☐ If fees are collected, has the system been tested to ensure the rapid exiting of vehicles?

☐ Have the standpipe systems been inspected and tagged? Note: Standpipe upper termini should be provided with approved pressure gauges, with double-drain valve assemblies.

☐ Are catch basins and other portions of the storm water drainage free of trash and litter? Note: Storm drains or sewer systems should have sediment traps installed.

☐ Are areas subject to erosion stabilized with grass, mulch, or other appropriate sediment control measures?

- Cooking and dining facilities:

☐ Have required gas pressure/leak tests been conducted and approved?

☐ Do all kitchen vent hoods have approved vent hood systems and vent hood fire suppression systems installed? Note: Generally, vent hood fire suppression systems require semi-annual inspections and tagging.

☐ Are there approved fire extinguishers in kitchens and dining areas?

☐ Are storage height clearances in storage areas as per code? Note: Generally, areas with sprinklers permit 18-inches clearances and areas with no-sprinklers 24-inches clearance.

☐ Are walk-in freezer/chillers without electrical shock hazards? Note: Common hazards are exposed wiring and bulbs.

☐ Are all electrical breaker panels and equipment unobstructed?

☐ Are fire alarm systems per code and are they approved and tagged by the local FD?

☐ Are exit signs that are installed for the kitchen staff per code?

- Main lobby, exit discharge level:

☐ Are all exit doors properly signed?

☐ Have all electronic egress control devices been approved and permitted?

☐ Are all assembly and discharge areas been approved and permitted?

☐ Are exiting areas properly lighted?

☐ Are required Fire Command Centers provided? Notes: (1) Fire alarm systems should be approved and tagged. (2) Where required, fire hoses and extinguishers should be provided and identified.

- □ Are light guard/tube protectors installed?
- □ Are elevators in working order and do they have fire service recall installed?
- Basement, below grade, and windowless levels:
  - □ Are stairway barriers in place to prevent people from running down into basements/cellars?
  - □ Are areas below ground fully sprinkled?
  - □ Are below grade–level exiting routes unobstructed and as per code? Note: Do not overlook below grade–level areas. There should be adequate exit signage and lighting.
  - □ Are all combustibles stored properly? Notes: (1) Check quantities to ensure allowable amounts are not exceeded; they may not be allowed at all. (2) Check heights, aisle widths, and lighting; guards/tube guards should be installed.
  - □ Are fire hoses and extinguishes provided, identified, and tagged?
  - □ Are trash, waste, and recycles containers stored in approved locations?

MECHANICAL

- HVAC: See above checklists pertaining to HVAC.
  - □ Have temperature controls been operated and tested?
  - □ Are fixtures/equipment properly adjusted and clean?
  - □ Is all equipment leak free?
  - □ Have all specified accessories been installed?
  - □ Have all equipment/appliances been tested and are there records of the tests?
  - □ Have cleanouts, decorative cover plates, and so forth, been checked?
  - □ Are floor drains connected to the storm sewer system? Note: Pour water into the drains to ensure that they do indeed drain.
  - □ Have registers been properly adjusted?
  - □ Have all controls been tested under operational conditions?
  - □ Are flues, vents, and so forth, per code?
  - □ Are all manufacturer's instructions, data, and so forth, available to staff?
- Central plant, maintenance service areas: Prior to acceptance, qualified operators should operate all equipment; if training is required, this should have been provided prior to acceptance. Equipment should be clean and leak free. A preventive program for all equipment should be in place in such a way that careful, accurate, and complete records can be kept. If required, special tools should be on hand and tools should be on well-planned tool board or toolboxes and not on benches or in drawers. There should be an inventory of spare parts and special supplies such as lubricants and cleaning supplies.
  - □ Have all standpipe and sprinkler system(s) risers and valves been properly inspected and tagged?
  - □ Have all fire pumps[s] been properly inspected and tagged?
  - □ Have boilers been permitted to operate? Note: Check area for oil, steam and water leaks.
  - □ Are all flammable and combustible liquids properly stored? Note: All hazardous materials should be identified and stored. Besides those already

mentioned, check such things as chemicals, gas cylinders, and so forth. All areas should be clean and orderly.

☐ Has all noncombustible storage in boiler, furnace, and mechanical rooms been identified, inspected, and service tagged?

☐ Have all special fire suppression systems been inspected and tagged?

☐ Are all fire alarm systems operational and tagged?

☐ Are all required permits on hand?

☐ Are emergency generators operational and Fire Department approved?

☐ Is all piping as per code?

☐ Has Fire Department issued all required permits?

☐ Are floor drains connected to the sanitary sewer system?

ELECTRICAL

Prior to a final inspection, all electrical work would have been subjected to a rough-in inspection that was done prior to insulation having been installed and the walls being closed up. The electrical inspectors had an opportunity to clearly see the important work. However, a second, final inspection is generally required when the building is complete, but before the building can be occupied. At this point, all of the walls are closed in, paint is finished, floors are complete and the owners are ready to install furniture and therefore this inspection covers only that which is visible.

- Are all circuits functioning and has every light been hung properly?
- Are outlet and switch heights consistent? Notes: (1) Check receptacle and switch heights to ensure code and ADA compliance.
- Is all wiring properly anchored? Notes: (1) Wires should be attached to wall studs to secure them. The first staple should be no farther than 8 inches from a box and then at least every 4 feet thereafter. (2) Cables should be run through the center of wall studs to help keep the wire from drywall screws and nails. The horizontal runs should be at least 20–24 inches above the floor and a metal wire–protective plate should protect each wall stud penetration.
- Are power taps (strips) used in an approved manner?
- Are electrical wall outlets, switch cover plates, control boxes, and electrical extension cords used in an approved manner?
- Does all emergency power and lighting comply with specifications and code?
- Are walk-in freezers and chillers properly grounded? Note: There should be no exposed wiring or bulbs.
- Has the exterior security lighting been inspected at night?
- Are electrical breaker panels and equipment unobstructed?
- Are service systems, for example, intercom, cable, IT, fire alarm, and so forth properly labeled?

ELEVATORS

Almost everywhere elevators must be inspected and authorized for use by a city/county/state inspector or one who is licensed by one of these entities. The checklist that follows was extracted from checklists used by the Elevator Division of New York City's Department of Buildings (hydraulic, machine roomless (MRL), and traction). While you will probably not be the person who conducts the official inspection, the checklist that is provided should help you prepare for it. Prior to an inspector arriving, you should record all device information, for example, device number, rated speed, capacity (pounds), manufacturer, code data plate, and so forth.

At the final building inspection, make sure that all of the elevators have been properly inspected and that their permits are posted.

- Do the following comply with the local elevator code and has an elevator inspector approved them?
- Inside of cars—all elevators:
    - ▫ Door reopening device?
    - ▫ Stop switch?
    - ▫ Operating control devices?
    - ▫ Car floor and landing sills?
    - ▫ Car lighting and receptacles
    - ▫ Car emergency signals?
    - ▫ Door closing forces?
    - ▫ Power closing of doors/gates?
    - ▫ Power opening of doors/gates?
    - ▫ Car vision panels/glass doors?
    - ▫ Car enclosures: width, depth?
    - ▫ Emergency exits?
    - ▫ Ventilation?
    - ▫ Signs/operating symbols? Braille?
    - ▫ Standby power operation?
    - ▫ Restricted openings?
    - ▫ Car rides?
- Machine rooms—traction and hydraulic elevators:
    - ▫ Access to machinery/spaces?
    - ▫ Headroom/lighting, receptacles?
    - ▫ Housekeeping?
    - ▫ Ventilation?
    - ▫ Fire extinguishers?
    - ▫ Controller wiring, fuses, and groundings?
    - ▫ Exposed equipment/guards?
    - ▫ Disconnecting means and controls?
    - ▫ Governors, over-speed switches, and seals?
- Machine rooms—traction elevators:
    - ▫ Static controls?
    - ▫ Overhead beams and fastenings?
    - ▫ Drive machine brake?
    - ▫ Traction drive machines?
    - ▫ Gears, bearings, and flexible couplings?
    - ▫ Winding drum machines and slack cable devices?
    - ▫ Belt- or chain-drive machines?
    - ▫ Motor generators?
    - ▫ Absorption of regenerated power?
    - ▫ Traction sheaves?

- ☐ Secondary and deflector sheaves?
- ☐ Rope fastenings?
- ☐ Terminal stopping devices?
- ☐ Car/counterweight safeties?
- ☐ Low oil protections?
- ☐ Static controls?
- ☐ AC drives from DC source? (Usually found in old elevators only)
- Machine rooms—hydraulic elevators:
  - ☐ Control valve tanks?
  - ☐ Flexible hydraulic hoses and fittings?
  - ☐ Supply lines and shut-off valves?
  - ☐ Hydraulic cylinders?
  - ☐ Pressure switches?
  - ☐ Winding drum machines?
  - ☐ Hydraulic power units?
- Top of cars—hydraulic elevators:
  - ☐ Stop switches?
  - ☐ Car top lights and receptacles?
  - ☐ Top of car operating devices?
  - ☐ Clearance, refuge space, and standard railing?
  - ☐ Normal terminal stopping devices?
  - ☐ Final and emergency terminal/limiting devices?
  - ☐ Car leveling and anti-creep devices?
  - ☐ Top emergency exits?
  - ☐ Floor and emergency IDs?
  - ☐ Hoistway construction?
  - ☐ Hoistway smoke controls?
  - ☐ Pipes, wiring, ducts are per code?
  - ☐ Projections, recesses, and setbacks?
  - ☐ Hoistway clearances?
  - ☐ Multiple hoistways? Per code?
  - ☐ Hoistway door and gate equipment?
  - ☐ Traveling cables and junction boxes?
  - ☐ Car frames and stiles?
  - ☐ Guide rail fastenings and equipment?
  - ☐ Governor ropes?
  - ☐ Wire rope fastenings and equipment?
  - ☐ Suspension ropes?
  - ☐ Top counterweight clearances?
  - ☐ Car, overheads, and deflector sheaves?
  - ☐ Broken ropes, chains, and tape switches?
  - ☐ Counterweights and counterweight buffers?
  - ☐ Counterweight safeties?
  - ☐ Compensating ropes and chains?
  - ☐ Ascending car overspeed protections?
  - ☐ Unintended car motion?

- Top of cars—MRL elevators:
    - Gears and bearings?
    - Belt or chain drive machines? Record.
    - Overhead beams and fastenings?
    - Drive machine brakes?
    - Rope fastenings?
    - Governor, overspeed switches?
    - Car locking devices?
    - Locking brackets?
- Outside hoistways—MRL elevators:
    - Car platform guards?
    - Hoistway doors?
    - Vision panels?
    - Hoistway door locking devices?
    - Accesses to hoistways?
    - Power closing of hoistway doors?
    - Sequence of operations?
    - Hoistway enclosures?
    - Elevator parking devices?
    - Emergency doors in blind hoistways?
    - Separate counterweight hoistways?
    - Standby power select switch/panel
    - Inspection controls?
- Pits:
    - Pit accesses: lighting and light switches, receptacles?
    - Bottom clearance, run bys, and minimum refuge spaces?
    - Buffer and emergency terminal speed limiting devices?
    - Final and emergency terminal stopping devices?
    - Normal terminal stopping devices?
    - Plunger and cylinder (hydraulic elevator)?
    - Traveling cables?
    - Compensating chains, ropes, and sheaves?
    - Governor rope tension devices?
    - Car and frame platform?
    - Car safeties and guiding members?
    - Car buffers?
    - Supply piping?
- Firefighters' service:
    - Smoke detector recalls?
    - Phase I key switch operation?
    - Phase II key switch operation?
- ADA: See also Due Diligence Inspections—Larger Buildings, Limited Disabled Review.
    - Do all elevators provide space for wheelchair users to enter, maneuver within, reach controls, and exit the cars? Notes: (1) Doors should provide ≥36-inches clear openings. (2) Cab depths should be ≥51 inches, with

≥54 inches from rear of cabs to inside face of doors. (3) Cab widths for single-speed doors should be ≥68 inches and for center opening doors ≥80 inches.

☐ Are clearances between car platform sills and edge of hoistway landings ≤1¼ inches?

☐ Are elevator operations automatic?

☐ Are floor surfaces firm, stable, and slip resistant? Note: If carpet is used, it should have the following features: be securely attached; have a firm cushion, pad, or backing (or none); have a level loop, textured loop, level-cut pile, or level-cut/uncut pile texture; have a pile thickness of ≤½ inches; and have all exposed edges fastened to the floor surfaces with carpet edge trim.

☐ Do cars have self-leveling features that bring the cars to floor landings within a tolerance of ½ inches underrated loading and zero loading conditions?

☐ Are hall call buttons centered at 42 inches above floor? Notes: (1) Hall call buttons should have visual signals to indicate when call is registered and answered. (2) The minimum size of hall call buttons should be ≥¾ inches in the smallest dimension. Notes: (1) "UP" buttons should be above "DOWN" buttons. (2) Buttons should be raised or flush. (3) Objects mounted beneath hall call buttons should not project more than 4 inches from the wall.

☐ Are car controls and car position indicator buttons ≥¾ inches in their least dimension? Note: Buttons should be raised or flush.

☐ Are control buttons designated by Braille and by raised standard alphabet characters for letters, Arabic symbols for numerals, or standard symbols as required in ASME A17.1? Notes: (1) Controls should be located on a front wall if cars have center opening doors, and at either a sidewall or the front wall if cars have side-opening doors. (2) Characters should be ≥5/8 inches to ≤2 inches high, raised 1/32 inches, upper case, sans serif, or simple serif type, and shall be accompanied by Grade 2 Braille. (3) All raised designations should be immediately left of the button to which they apply. (4) Floor buttons should be provided with visual signals that light when each call is registered and extinguish when each call is answered. (5) All floor buttons should be ≤54 inches above the floor where side approach is provided, and ≤48 inches where forward approach is required. (6) Emergency controls, including alarm and stop, should be grouped at the bottom of the control panel, with center lines ≥35 inches above the floor. (7) A visual car position indicator should be above the car control panel or above the door. As the car passes or stops at a floor, the corresponding floor numbers should illuminate and an audible signal should sound. Numerals should be ≥ ½ inches high.

☐ Are visible and audible signals provided at each hoistway entrance to indicate which car is answering a call? Notes: (1) Audible signals should sound once for UP, twice for DOWN, or shall have verbal annunciators that say "UP" or "DOWN." (2) Audible signal should be ≥20 decibels with frequency ≤1500 Hz.

- [ ] Are visible fixtures mounted with centerline at ≥72 inches above the lobby floor?
- [ ] Are visual elements at least 2½ inches in the smallest dimension?
- [ ] Are signals visible from the vicinity of the hall call button?
- [ ] Do all elevator hoistway entrances have raised and Braille floor designations provided on both jambs? Notes: (1) Characters should be centered 60 inches above finish floors. (2) Characters should be 2-inches high, raised 1/32-inch upper case, sans serif or simple serif type, and should be accompanied by Grade 2 Braille.
- [ ] Do all elevator doors open and close automatically?
- [ ] Does each door have a reopening device that stops and reopens whenever the door becomes obstructed by an object or person? Notes: (1) The devices should be capable of completing these operations without requiring contact when the obstruction passing through the opening is ≥5 inches and ≤29 inches above finish floor. (2) Door reopening devices should remain effective for at least 20 seconds; after such interval, doors may close in accordance with ASME A17.1.
- [ ] Are the minimum times from notification that a car is answering a call until the doors begin to close as follows:

$$T = D/(1.5 \text{ ft/s})$$

where:

T = time in seconds

D = The distance from a point 60 inches directly in front of the furthest call button to the centerline of hoistway door.

Notes: (1) The notification time should be ≥5 seconds for hall calls. (2) The time for elevator doors to remain fully open in response to a car call should be ≥3 seconds.

- [ ] When emergency two-way communication systems between the elevator and a point outside the hoistway are provided, do they comply with ASME A17.1? Notes: (1) Highest operable part of each system should be ≤48 inches from floor. (2) Systems should be identified by raised symbols and lettering located adjacent to the device they serve. (3) Characters should be ≥5/8-inches to ≤2 inches high, raised 1/32-inches uppercase, sans serif or simple serif type, and should be accompanied by Grade 2 Braille. (4) If a system uses a handset, cord length should be ≥29 inches. (5) If located in a closed compartment, the door should be operable with one hand, should not require tight grasping, pinching, or twisting of the wrist, and should require a maximum force of 5 lbf.[37] (6) The emergency communication system should not require voice communication. (Voice-only systems are inaccessible to persons with speech or hearing impairments.)

---

[37] lbf: A nontechnical unit of force equal to the mass of 1 pound with an acceleration of free fall equal to 32 feet/sec/sec.

PART 2

DUE DILIGENCE[1] INSPECTIONS, AND EXISTING BUILDINGS CAPITAL PROJECT PLANNING

Here, due diligence inspections and existing building capital project planning inspections are lumped together because the information being sought is essentially the same. The difference being that the due diligence inspection is for a prospective buyer, whereas the existing building capital project planning inspection is for an owner who wants to maintain and perhaps improve his/her facility.

The section on house/light construction speaks almost exclusively to due diligence because, while people may plan to maintain and improve their homes and small mini-malls, most people don't have a formally drafted long-range plan; it may, however, be in their head. But people who intend to purchase a home or mini-mall do want a report that tells them just what they are buying, and the mortgage bank may want such a report too.

In the case of larger buildings whose value may be many millions of dollars, prioritized plans are in order. So, at the risk of repetition, whether you are inspecting

---

[1] As used here, due diligence is a general measure of prudence, responsibility, and diligence that is expected from, and ordinarily exercised by, a reasonable and prudent person under the circumstances. For example, it is common to have a "due diligence" inspection conducted prior to purchasing a building.

for an owner or a potential owner, you want not only to know the condition of the facilities but probably will be asked to develop a plan of capital expenditure that goes into the future; the intent of the section on larger buildings is to help you accomplish this planning.

3

HOUSE/LIGHT CONSTRUCTION

Despite laws that are designed to protect homebuyers and smaller property owners, as we saw in the real estate crash of 2008, this group is in many ways at the bottom of the food chain. And, it is not only financial shenanigans that can get them into trouble. Most, especially first-time buyers, but even people who have purchased several properties, simply do not have the experience necessary to conclude that the building that they want to buy is sound. Recognizing this, the potential buyer asks his/her agent for the name of an inspector that the agent then provides. At minimum, this should be a certified inspector; it is usual for inspectors to be certified by an organization, not a state licensing board.[2] Therefore, I have somewhat changed the message of this section: rather than aiming it at those who perform the due diligence inspections only, it is also aimed at prospective buyers, the receivers of the inspection reports; the word you still refers to you, the inspector.

Contracts for house inspections usually require a written real estate inspection report that will identify any material defects that are discovered during the inspection and that in your opinion are safety hazards, things that are not functioning properly or appear to be at the end of their service lives. You are usually tasked with providing recommendations for correction or further evaluation. Nowadays, with the ease of inserting photographs and sketches into a report, most reports should include both, especially of deficiencies and potential problem areas.

The checklists that are provided for light construction in the first section of this book cover the same ground as do due diligence inspections and, therefore, can be used by you and your client when preparing for the due diligence inspection. The checklists give an understanding of the standards of construction that should be expected in the building that is being inspected.

---

[2] For example, the International Association of Certified Home Inspectors (InterNACHI is an international trade organization that is comprised mainly of independent inspectors. InterNACHI provides inspection courses and requires its members to complete 24 hours of continuing education annually. There are also state-specific organizations, for example, the California Real Estate Inspection Association (CREIA).

Earlier I mentioned that different formats, approaches if you will, to inspections are used in each major section of this book. In this section, I use one that is common for house inspectors. A common table of contents for the technical report that the future buyer will receive is as follows;[3] this may vary from area-to-area and inspector-to-inspector (there are also alerts for owners that need to be considered when hiring an inspector):

- **Definitions and scope:** Here the contractual limitations of the inspection are repeated, disclaimers added, and there may be some additional advise to the buyer about contacting an architect, engineer, or contractor for wherever there is concern.
- **Foundations, basements, and underfloor areas:** You should expect to provide information about the following: foundation system; floor framing system, including piers and cripple walls,[4] foundation and cripple wall anchoring (especially important in earthquake areas); and insulation if it exists. Deterioration and rot should be reported. It is important to note whether you actually entered the crawl space or if you considered the crawl space to have fallen under the "difficult to reach" exemption.

  You are not expected to evaluate the adequacy of the design, for example, floor joist spacing, or rate the adequacy of the insulation. However, should you notice a serious deficiency or uncommon practice, for example, an extra-wide spacing of floor joists, this should be reported.
- **Exteriors:** You should provide information on the following: whether the surface grading around the house is proper; the general condition of windows and doors; decks, porches exterior stairs; the exterior building envelope; and, the driveway and sidewalks.

  Your contract will normally not require you to climb a ladder to check carefully under the eaves; unfortunately, this is where there may be rot and sometimes wasp's nests. So, a pair of binoculars is handy and should be used. You are also probably not required to inspect fence posts where there may also be hidden rot, but a slight push will tell immediately if the post is rotten.
- **Roof coverings:** There are many types of roofs. By contract, you may have to do only a visual inspection from the ground. However, if the house has a flat roof, the only way to see it is by climbing a ladder. Make sure that you and your client are clear on this issue. If you go onto the roof, you should report on the condition of the covering, drainage, especially signs of ponding and the buildup of debris; the condition of the flashing, for example, what flashing materials are present, for example, copper, aluminum, galvanized metal, whether there is a danger of different flashing materials corroding due to galvanic action, and if the flashing has been installed properly; the condition of penetrations, for example, kitchen and bathroom vents and skylights; and,

---

[3] Much of the information on home inspections comes from the California Real Estate Inspection Association's (CREIA) Standards of Practice. Generally, CREIA's Standards of Practice follow closely those of other state inspection associations.

[4] Stud walls that are built on top of an exterior foundation to support a house and create a crawl space are called cripple walls. They carry the weight of the house. During an earthquake, these walls are subject to collapse if not properly braced to resist horizontal movement.

other problems such as birds nesting.

Where the roofs are visible from the ground, you will normally conduct your inspections from ground level or perhaps the eaves. Report all active roof leaks, but unless it is raining when you conduct your inspection, do not guarantee that the roof will not leak. If you venture a guess as to the covering's age, do it in such a way that you cannot be held liable for the estimate.

- **Attic areas and roof framing:** This is another area that may be "difficult to reach" and thus seen only from afar. You will usually be able to comment on the adequacy of attic ventilation (important to prevent rot); the condition of the framing, for example, the rafters of older houses may be sagging because they have spans that are longer than currently allowed; and, the general condition of the insulation. Your report should certainly tell if you spotted any traces of vermin.
- **Plumbing:** While the piping is probably hidden, your report should cover the following: whether the water piping is copper (if possible); the location, type, and condition of the main water shut-off valve and the water supply piping; sewage piping and vent pipes (to the extent that they are visible); faucets and fixtures, including anti-siphon devices or a comment that there are none; rates of flow into and out of sinks; whether toilets flush properly; the location and condition of the gas shutoff, and in earthquake areas that there is/is not an earthquake emergency shutoff. You should report on the condition and type of the water heater[s] and whether it has a sediment trap (most manufacturers require one). Also, report on whether the water heater[s] is properly braced, that the overflow is properly piped, and that it is properly installed.

  If you are using a standard house inspection contract, you probably will not be required to fill fixtures with water or check overflow drains or drain-stops; evaluate backflow devices, waste ejectors, sump pumps, and drain line cleanouts or solar heating systems or components; inspect or evaluate water temperature balancing devices and temperature fluctuations; clock the time it takes for hot water to arrive at a sink; inspect such items as whirlpool baths, steam showers, or sauna systems; inspect fuel tanks for sludge or gas lines for leaks; and, inspect wells and water treatment systems. These are all fairly specialized items. If your client wants them inspected and evaluated, he/she will have to have them inspected separately.
- **Electrical:** As with plumbing, most of the wiring will be hidden, but unless deliberate fraud is involved, one can get a pretty good idea of the condition of the electrical system. Your report should cover the following: the main electrical panel and its amperage (nowadays many codes allow 100 amps to a residence, but 200 amps are much preferable) and subpanels; circuit wiring including the wiring to any detached buildings, for example, a garage; switches, receptacles, outlets, and lighting fixtures, paying special attention to whether ground fault circuit interrupters (GFCI) are installed per code, for example, kitchens, exterior outlets, garage conversions, and bathrooms and whether receptacles are wired properly and do not have reverse polarity.

  You will normally not be required to operate circuit breakers or circuit interrupters; remove cover plates; inspect de-icing systems or components; or inspect private or emergency electrical supply systems or components.

- **Heating, ventilation, and cooling (HVAC):** The following HVAC equipment should be checked: central cooling equipment; energy source and connections; combustion air and exhaust vent system; condensate drainage; and conditioned air distribution systems. This usually means operating the equipment from the thermostat and conducting a visual inspection without removing cover plates or checking airflows with testing equipment.

  You will normally not inspect heat exchangers or electric heating elements; inspect noncentral, for example, window air conditioning units or evaporative coolers; inspect radiant, solar, hydronic,[5] or geothermal systems or components; determine volume, uniformity, temperature, airflow, balance, or leakage of any air distribution system; or inspect electronic air filtering or humidity control systems or components.

- **Fireplaces and chimneys:** Some house inspection contracts exclude fireplaces and chimneys; again, make sure your client knows whether fireplaces and chimneys are in or out. If they are included you need to ensure that the chimney is in compliance with earthquake codes; these codes are often retroactive, in which case the seller would be held responsible for any necessary upgrade. If the chimney is on the exterior of the house, you should make certain that the fireplace has not shifted away from the house (fireplaces are usually built on their own foundations that sometimes tilt) and that the cricket[6] is in good condition. Where the chimney comes through the roof, you should make sure that the flashing is free of corrosion and debris and is watertight. Check the spark arrestor, bird screen, and rain cover to ensure that they are properly installed. Check to ensure that the damper closes and opens smoothly, that there is no dangerous buildup of creosote, and other byproducts of burning, and that the hearth extension is tight and will not allow embers to fall through.

  Usually you will not be required to operate the fireplace, nor will you usually have the tools and equipment to ensure that there are no cracks in the masonry, burn out holes in metal "insert" type fireplace boxes, or that all seals and gaskets are totally safe. Each of these deficiencies may allow flue gasses to enter the house, so don't be a hero and get in over your head. When speaking with your client, give him/her a warning: There are firms that advertise themselves as fireplace experts when they really aren't. Your client should carefully select any fireplace expert who is hired.

- **Building interior:** You should visually inspect the entire interior of a house: walls, ceilings, and floors, looking for such things as cracks, sagging ceilings, sloped floors (place a ball on the floor), and loose floorboards; doors and windows, checking especially that they do not get hung up due to settlement, that locks are in place and work, that window frames are not rotted or letting water enter the building envelope, and so forth; stairways, looking especially at clearances and tripping hazards and proper lighting; handrails and guardrails, looking for code compliance and soundness; permanently installed appliances,

---

[5] Hydronic: Of, relating to, or being a system of heating or cooling that involves transfer of heat by a circulating fluid (as water or vapor) in a closed system of pipes.

[6] Cricket: As used here, a saddle-shaped, peaked construction that connects a sloping roof with a chimney. Crickets are designed to encourage water drainage away from the chimney joint.

generally checking whether they function and if there are safety hazards such as loose wiring; smoke and carbon monoxide alarms, checking to see that they are present; and garage doors.

Normally, you will not be required to operate smoke and carbon monoxide alarms, determine whether security is adequate, or climb ladders to check components and systems.

- **Glossary of terms:** Glossaries, if included, normally include words and terms that apply to home ownership and the inspection, for example, functional flow (the flow of the water supply at the highest and farthest *fixture* from the *building* supply shutoff valve when another *fixture* is used simultaneously).

- **List of limitations, exclusions, and not-required-to's:** Your contract should explain that the purpose of the inspection is to provide information about the general condition of the building(s). To accomplish this, you will conduct a survey and check on the basic operation of the systems and components of the building. Under a standard contract, you will normally limit 7 your inspection. You will not identify concealed or latent defects; determine whether the property is suitable for its intended use; comment on market value or insurability; predict life expectancy of the property or components; get behind permanently installed items; and comment on cosmetic and esthetic conditions.

Your standard contracts will normally exclude you from making a determination on: property line and encroachment matters; components and systems that cannot be reached, entered, or viewed without difficulty; service life expectancy of components or systems; the size, capacity BTU, and efficiency of components and systems, although newer equipment may have this information affixed to it; what may have caused a condition to exist; code compliance; the presence of evidence of rodents, birds, animals, insects, and so forth, unless obvious; the presence of mold, mildew, or fungus unless obvious; the presence of airborne/invisible hazards, such as radon, electromagnetic fields, and other forms of air pollution; the existence of environmental hazards such as lead paint, asbestos or toxic drywall; hazardous waste conditions; manufacturers' recalls or conformance with manufacturers' installation procedures (especially important with appliances such as dishwashers); adverse comments that may have been published by EPA, OSHA, and so forth; and acoustical properties. You generally should not venture to estimate costs to replace items or operate them.

Your standard contracts will state that you are not required to operate phone and cable lines, satellite dishes and antennas, lights, intercoms, speaker systems, security systems, and remote controls; systems that do not turn on with the use of normal operating controls; shut-off and manual stop valves; electrical disconnects and overcurrent protection devices; alarm systems; and moisture meters, gas detectors, or similar equipment.

Your standard contract will also say that you are not required to move personal items or other obstructions such as ceiling tiles, ice, debris, or anything else that might get in the way of a visual inspection; dismantle,

---

[7] The limitations, exclusions and list of items that inspector are not required to perform were taken from International Standards of Practice for Performing a General Home Inspection, that was revised June 2013.

open, or uncover any system or component; enter an area deemed unsafe or do anything deemed unsafe by the inspector (this may include such things as entering a crawl space, walking on a roof, and climbing a ladder); inspect underground items such as irrigation systems and underground storage tanks; inspect decorative items; inspect common elements of a multi-unit house; offer guarantee and warranties; research property records; and determine the age of the building or its components or to differentiate between original construction and subsequent alterations, additions, and replacements (On older houses, this may mean the difference between 2×4s that measure 2 inches by 4 inches versus new 2×4s that measure 1.5 inches $\times$ 3.5 inches). Unless you are a licensed engineer, you should not offer professional engineering services such as structural analyses and environmental audits.

The bottom line when it comes to most home inspections is that you can offer a potential buyer a sense of the general condition of the house, but your client should understand that your inspection is far from all inclusive. For example, besides the **limitations, exclusions,** and **not-required-to's** already noted above, a seller may show that the house is termite free based on recent termite eradication and a certificate from the licensed contractor, but because termite damage is hidden, a standard home inspection would not reveal the extent of such prior damage, which can be considerable. If you have the slightest suspicion that there is termite damage or mold, you should ask to look behind the plaster or gypsum board walls or warn that this should be done. Similarly, if you have any suspicion that sewer pipes may be broken, a camera inspection may be in order. Needless to say, each additional inspection adds cost.

4
LARGER BUILDINGS

In this section I am discussing "larger buildings." Because there are so many variations of larger buildings, for example, large office buildings, factories, academic buildings, and on-and-on, I found it impossible to be more specific. Whatever the size, age, and type of facility, it will represent a sizeable investment. A potential buyer will want to know what he/she is getting into and, therefore, will want a thorough inspection of the building(s) and its ancillary infrastructure. The inspection should evaluate functional performance, identify and prioritize needed work, and usually attach some estimates of cost. Such an inspection would be guided by the nature of the facility. It may also be guided by how the owner wishes to receive the inspection report. For example, the owner's budgeting methods and organizational structure may guide your report's formatting so that it is in sync with his/her maintenance work order system and capital budgeting and planning process. Depending on the size and complexity of the building, such a report could be quite complex and costly. If the owner adds a complete records search to the technical inspections, it becomes even more costly.

Therefore, the checklists provided in this section are somewhat general. You also have the more detailed checklists that were provided earlier. I am assuming that you have knowledge of facilities and will be able to take the information of this and previous sections and build upon them. As with the other sections of this book, "you" refers to you the inspector. If instead you are an owner, this next section should provide information on what to demand from your inspection team.[1,2]

Due diligence inspections and inspections intended to develop capital plans for major maintenance and repairs of existing facilities provide a record of deficiencies and estimate of costs to correct them. The condition inspection part of the process is a visual and sometimes invasive inspection of buildings and infrastructure. Inspections usually begin with the grounds upon which the buildings are constructed; how the building is placed in the ground and then they move upward to include structural framing, the building envelope and interiors, taking into account utilities, vertical transportation,

---

[1] Again, a different format is presented for your consideration.
[2] Not included here are distributed utilities, for example, steam and on-site generated electricity, to multiple buildings.

and safety; some inspectors prefer the reverse order, starting at the roof and working down. What is important is to organize your inspection so that nothing is missed. Because of the complexity of such facilities, different people will probably inspect each service system separately. A comprehensive inspection includes architectural, civil/ structural, mechanical, electrical, and safety components. A typical Table of Contents for the report being discussed here is as follows:

- Site
- Structural/seismic
- Building exterior
- Roofing
- Building interior
- Disabled access review
- Heating, ventilating, and air conditioning systems (HVAC)
- Plumbing
- Electrical
- Fire and safety
- Vertical transportation

Costing to correct deficiencies would typically be done using commercial estimating guides, unless the owner provides this information.

- **Planning the inspection:** An inspection's scope is determined by the client's intended use of the inspection results. In considering the scope of an inspection, take into account the following: the specific facilities to be inspected; intended use of results (one time or to build on future inspections); current available information; and established deadlines and access to facilities. In some areas, weather conditions need to be considered, for example, it is not really possible to inspect a roof that is covered with snow. Procedures should be established to promptly correct unsafe/emergency conditions.
- **Scheduling the inspection:** The timetable for conducting the inspection establishes staffing needs. Factors such as the facility's size and age, access, available information, and the amount of support from in-house personnel all play a role. Unless you are using a computerized system as is described in the next section, inspecting a School District, a Campus, or similar, you will find that it takes considerable time to prepare the report, that will probably look very much like a manual. Building occupants and users of the infrastructure should be notified of the inspection schedule.
- **Conducting the inspection:** Individual component inspections are usually summarized on inspection report forms. Inspectors are typically responsible for recording deficiencies, prioritizing them, and estimating the cost for correcting them. Sometimes engineering support will be required if, say, a major HVAC component must be replaced.
- **Priority rating:** Priorities are based on the importance of the necessary corrective action. The priority rating system sorts out the relative importance of each action such as preventing further loss, life safety, meeting code, and conserving energy. Many owners have established their own system of prioritizing. The exact system to be used should be established prior to the start

of the inspection. Here is one approach to prioritizing.[3] Another approach is provided in the following section on schools:

- ☐ Urgent/unsafe (Priority 1): Items identified as unsafe need to be repaired immediately. Urgent repairs should be made within one year. These normally include items which, if not corrected will lead to further and rapid deterioration.
- ☐ Safe with repair and maintenance (Priority 2): These are deficiencies that if not corrected within approximately two years will become "Urgent." Also, quick payback projects to reduce energy and operating costs are classified in this category.
- • Safe (Priority 3): Deficiencies that do not fit into the previous two categories but which, if deferred longer than three to five years will affect the use of the facility or cause damage to it. Roofs that are showing some wear would fall into this category.

You may also wish to build into your prioritization plan consideration of the way facilities deteriorate and costs escalate (generally ess-shaped curves in the opposite direction). Generally, a facility or a component starts to deteriorate slowly. Take a small tear in a roof that was caused by hail. It may not even leak, but if ignored it will start leaking. Then the tear starts to get bigger and bigger. And then, before you know it, you must make major repairs to the roof and to the building below; for want of a nail, the war was lost. So, while the purposes of the inspections we are discussing are due diligence and capital planning, you may wish to include the establishment of a large-scale preventive maintenance programs in your capital plan budget.

- • **Reports:** Most owners who are planning to hire a firm to conduct their due diligence/capital planning inspection will ask that you show them previous reports that you have prepared. Such reports would, as a minimum, normally include an executive summary that provides an overview and highlights findings; a narrative that provides a complete description of the facility; a needs assessment of maintenance and capital improvements; and an assessment of the facility in light of current "green" and sustainability and Americans with Disabilities Act (ADA) concerns. Your reports should answer the dual questions: Is the facility suitable for its intended use, or will it require extensive remodeling; and, what will the cost be to bring the building up to the desired standard (compared to a new building or relocation to another site)? In addition, it can be the impetus for starting a routine inspection and maintenance program. Reports are also helpful in producing feasibility studies of changes in building use that require alteration and renovation work.

It is important that there is understanding as to what will and won't be provided. All parties should be clear as to definitions and scope of the inspection and how it

---

[3] The priority rating system given here comes from New York City's facade ordinance, commonly referred to as Local Law 11 that is provided in Section 27-129 of the Building Code of the City of New York. Rule 32-03. LL-11 relates to the periodic inspection of exterior walls and appurtenances of buildings in New York City.

will be reported. Usually, this means a fairly detailed contract at the beginning of the process and a published report that includes your findings and estimates of costs at the end of the process.

SITE

The first impression one gets of a building or campus is on entering the site. Both design and maintenance leave a lasting first impression. Beyond appearances, site design and management issues affect the local ecology in general and immediate neighbors specifically. Improperly graded and poorly maintained grounds directly affect the value of the property and that of adjoining properties. Drainage patterns that cause water to enter neighboring properties are probably in violation of local building codes and will have to be corrected. Planting and landscape maintenance practices directly affect staffing and amount of other resources that must go into grounds keeping. Impervious surfaces contribute to urban heat island effect and cause harmful storm water runoff. Exterior lighting may cause unnecessary nighttime light pollution and unnecessary costs.

Sites may affect on a broader scale environmental concerns such as public transportation. For example, a large site that is near to public transportation may cause a nearby station to be constructed, thereby enabling employees to park their cars.

In developing the checklists that pertain to the site, I have used criteria of the U.S. Green Building Council (USGBC), Leadership in Energy and Environmental Design (LEED),[4] and environmental/sustainability requirements; owners will have to decide whether these are important to them; these criteria are constantly being upgraded so inspectors need to stay abreast of changes.

GRADING, DRAINAGE, AND LANDSCAPING

- Mitigated runoff:
  - □ The volume of runoff depends on perviousness of the surfaces and systems for capturing storm water and transporting it off site, be it to a wetland or urban infrastructure.
  - □ Identify major features of the site and compare the site to the as-built plans.
  - □ Runoff is defined as storm water that leaves the site via means of uncontrolled surface streams, rivers, drains, or sewers. LEED for existing buildings offers one point for implementing a storm water management plan that infiltrates, collects, and reuses runoff or evapotranspirates[5] runoff from at least 15% of the precipitation falling on the whole site both for an average weather year and for the 2 year, 24-hour design storm (always subject to redefinition). Mitigated storm water is equal to the volume of precipitation that falls on a site that does not become runoff.
  - □ If desired, you can calculate the percentage of mitigated runoff using the rational method: STET. The Rational Method refers uniquely to this calculation.

---

[4] LEED: Leadership in Energy & Environmental Design.
[5] Evapotranspiration: The return of water vapor to the atmosphere by evaporation from land and water surfaces and by the transpiration of vegetation.

$$Q = CiA$$

where, C = runoff coefficient
I = rainfall intensity (inches per hour)
A = drainage area (acres)

DEFAULT RUNOFF COEFFICIENTS

SURFACE TYPE	RUNOFF COEFFICIENT	SURFACE TYPE	RUNOFF COEFFICIENT
Pavement, asphalt	0.95	Turf, flat (0–1% slope)	0.25
Pavement, concrete	0.95	Turf average (1–3%)	0.35
Pavement, brick	0.85	Turf, hilly (3–10%)	0.40
Pavement, gravel	0.75	Turf, steep (>10%)	0.45
Roof, conventional	0.95	Vegetation, flat (0–1%)	0.10
Roof, garden roof (<4 in)	0.50	Vegetation, average (1–3%)	0.20
Roof, garden roof (4–8 in)	0.30	Vegetation, hilly (3–10%)	0.25
Roof, garden roof (9–20 in)	0.20	Vegetation, steep (>10%)	0.30
Roof, garden roof (>20 in)	0.10		

- Surface drainage:
  - ☐ Has an effort been made to reduce impervious surfaces, including construction of pervious walkways, and parking areas?
  - ☐ Is rainwater captured and reused?
  - ☐ Have filtering systems, bio-swales,[6] retention/detention basins,[7] channels, dikes, or vegetated filter strips or other uses of vegetation been installed to minimize runoff?
  - ☐ Are storm sewers protected from sedimentation?
  - ☐ Are manholes, inlets, and catch basins set to avoid ponding? Are they free of debris?
  - ☐ Has the land been contoured to direct storm water runoff through the site to give vegetation an additional water supply?
  - ☐ Have water retention and detention features been integrated into the overall site development? If required, calculate the volume captured via collection facilities.

---

[6] Bioswale: A landscaped element designed to remove silt and pollution from surface runoff. Swales as used here are shallow, trough-like depressions that carry water during rainstorms or snow melts. They have gently sloped sides (generally less than 6%) and are filled with vegetation, compost, and/or riprap.

[7] Retention/detention basin: A basin that is meant to collect storm water and slowly release it at a controlled rate. The main difference between a detention basin and a retention basin is whether it has a permanent pool of water; a retention basin is sometimes referred to as a wet pond, whereas a detention basin is sometimes referred to as a dry pond.

The volume of captured volume can be calculated using the following equation (extracted from USGBC, Existing Buildings: Operations and Maintenance (O&M) Reference Guide, August 2008. Source: 2000 Maryland Stormwater Design Manual, Vols. I, II (MDE, 2000)):

$$V_r \text{ (cubic feet)} = (P)(R_v)(A)/12"$$

where:
V_r = volume of captured runoff
P = average rainfall event (inches)
$R_v = 0.05 + (0.009)(I)$, where I = percent impervious of collection surface
A = area of collection surface (square feet)

Example: Rainwater is harvested from a 10,000-square foot roof (100% impervious). The system is designed to capture the runoff from 90% of the average rainfall event (1 inch of rainfall for humid watersheds). The volume of the proposed storage system is the amount of runoff captured (V_r):

$$V_r = (1\text{-})((0.05) + (0.009)(100))(10,000)/12$$

$V_r = (1)(0.95)(10,000)/12 = 791.67$ cf or 5,922 gal. (There are 7.48 gallons in a cubic foot.)

The tank must be emptied after each storm. Using a tank that is $10 \times 10 \times 8$-feet gives a total storage volume (V_s) of 800 cf. Using a design storm interval of three days (72 hours), the drawdown rate (Q_r) is

$$Q_r = 800 \text{ cf}/259,200 \text{ s} = 0.003 \text{ cfs or } 1.37 \text{ gpm}$$

The amount of runoff reduced by a storm water harvesting system is based on its storage volume, the rate at which the system is emptied, and the interval between storm events.

- Landscaping:
  - Does the site use a conventional (by local standards) irrigation system or does it have a system that uses high-efficiency technology, captured rain/recycled site water, or landscaping and other techniques to reduce potable water consumption?
  - In urban areas, have measures been taken to reduce the use of potable water for watering any roof/courtyard and garden space or indoor planters?
  - Have native plants been included in the site's landscaping? To what extent are they being used (percent of site)?
  - Have plants that yield berries, flowers, or leaves that can be tracked into building entrances been avoided?
  - Do trees shade nonroof impervious surfaces? Estimate the extent of shading.
  - Is there an off-site area with native or adaptive plants?
  - Are measures in place to prevent the loss of topsoil through wind erosion?

ROADS, WALKS, PARKING LOTS, AND OTHER PAVED SURFACES

See also High-Rise Building Construction, Site Work.

- Paved/impervious surfaces:
  - ☐ Has an inspection been made of all paved surfaces including alignment, surface, and subsurface conditions; settling and uplift; cracks and holes; and drainage and slope? Note: Inspectors should provide a description of all curbing: alignment, erosions, and needed repairs?
  - ☐ Do the alignments of truck routes, including turn-around dimensions, access to loading docks and parking meet the (new) owner's needs?
  - ☐ Has any attempt been made to capture and reuse storm water from impervious surfaces such as parking lots, roads, and roofs?
  - ☐ Have measures been taken to reduce heat islands (thermal gradient differences between developed and undeveloped areas)? Note: The reduction of heat islands minimizes adverse effects on microclimate and human and wildlife habitat?
  - ☐ What percent of impervious site areas, including parking lots, do trees shade? Note: Obviously an estimate on the inspector's part.
  - ☐ Have nonroof impervious surfaces been coated with light-colored/high-albedo[8] materials (reflectance of at least 0.3)?
  - ☐ Have pervious paving options been used, for example, porous concrete and asphalt or flexible plastic porous pavement?
- Parking:
  - ☐ Does the number of parking spaces comply with the local building code requirement?
  - ☐ Is striping easily visible?
  - ☐ Are sufficient Accessible Parking Spaces provided? Note: Comment on the location of Accessible Parking Spaces. (References: International Building Code, 2006, Section 1106, United States Department of Labor, ADA, and local ordinances.)
  - ☐ Is access from Accessible Parking Spaces to buildings unhindered?
  - ☐ Are 3% of the parking spaces dedicated to alternate fuel vehicles and/or car/van pool vehicles?
  - ☐ What percent of parking is underground or stacked?
  - ☐ What percent of outdoor parking areas use an open-grid pavement system?

REDUCTION OF ONE-PERSON AUTOMOBILE USE

- Building location vis-à-vis public transportation:
  - ☐ Is the building located within ½ mile of a commuter rail, light rail, or subway station?

---

[8] Albedo: The fraction of solar energy (shortwave radiation) reflected from the Earth back into space. It is a measure of the reflectivity of the earth's surface. Ice, especially with snow on top of it, has a high albedo; so a white roof that reflects the sunlight hitting has a high albedo, whereas a black roof that absorbs the sunlight has a low albedo.

       □   Is the building located within ¼ mile of two or more public bus lines?

       □   Are building occupants provided with a conveyance link to commuter rail, light rail, subway, or bus?

- Bicycles:
  - □ Is there nearby secure bicycle storage for 1% of the building's occupants?
  - □ Are showers provided for bicycle riders at the rate of one shower for each eight secure bicycle storage spaces?

SIGNAGE

Signs should be fresh and clear. Signs that have information painted on them, either in words or solidly painted sections that inform through color codes such as yellow and red to signify a fire lane should be easy to read and understand. Signs and sign support structures should be well maintained.

- **Site amenities:** Comment as appropriate.

STRUCTURAL/SEISMIC

Owners require an accurate assessment of the building's performance and capabilities. They need to know if it can be used for its intended purpose or if it will need costly and time-consuming rehabilitation and renovation. Owners do not want surprises such as having to meet more stringent building code requirements that are retroactive; for example a code may require the retrofitting of structural reinforcement. Also, changes in site conditions since the original construction may dictate the need for modifications to the structural systems. You should guard against over extending yourself; in some cases a structural engineer may be needed.

Reports must be based on an inspector's professional experience and judgment, which the inspector must apply as he/she examines the structural condition of the combinations of materials: concrete, metals, masonry, and wood.

STRUCTURAL CRITERIA

- Initial Screening Factors: Make a general assessment of the facilities, for example, general description paying attention to overall maintenance, number of stories, date of construction, dimensions, occupancy and usage, and so forth. You may also want to obtain the names of key facility staff members.
- Document Review: Review current, applicable building codes against the codes at the time of construction; in some cases you may have to review drawings and calculations. Determine if there are critical code differences on matters such as roof loads, floor loads, wind-loading criteria, level of seismicity, and allowable soil-bearing capacity. Check if the building department required geologic/soils reports prior to issuance of the building permit. Attempt to learn what materials were used, for example, concrete strengths in foundations and slabs.
- Building structural type: Identify the group into which the building best fits. The groups of "building types" listed below were first defined in ATC-14[9] and since then have been used in most of the FEMA guideline documents. Because most structures are unique in some fashion, judgment should be used where

---

[9] ATC: Applied Technology Council.

selecting the building type, with the focus on the lateral-force-resisting system and elements.[10]

- □ Wood light frames
- □ Wood light frames (multistory, multiunit residential)
- □ Wood frames, commercial, and industrial
- □ Steel moment frames (with stiff diaphragms)
- □ Steel braced frames (with flexible diaphragms)
- □ Steel braced frames (with stiff diaphragms)
- □ Steel light frames (with stiff diaphragms
- □ Steel frames with concrete shear walls
- □ Steel frames with infill masonry shear walls (with stiff diaphragms)
- □ Steel frames with infill masonry shear walls (with flexible diaphragms)
- □ Concrete moment frames
- □ Concrete frames with infill masonry shear walls (with stiff diaphragms)
- □ Concrete frames with infill masonry shear walls (with flexible diaphragms)
- □ Precast/tilt-up concrete shear walls (with flexible diaphragms)
- □ Precast/tilt-up concrete shear walls (with stiff diaphragms)
- □ Precast concrete frames (with shear walls)
- □ Precast concrete frames (without shear walls)
- □ Reinforced masonry bearing walls with flexible diaphragms
- □ Reinforced masonry bearing walls with stiff diaphragms
- □ Unreinforced masonry bearing walls with flexible diaphragms
- □ Unreinforced masonry bearing walls with stiff diaphragms

- Foundations: Comment especially on all elements that could affect seismic performance such as connections, anchorage conditions, the location of connections or supports, long and unsupported columns, undermining due to unknown water sources, and so forth.
- Building use: Determine if the building requires a higher than normal occupancy performance level, for example, a hospital. Look also for special architectural features such as finishes, registered historic features, and adjacent buildings that may cause pounding or falling hazards.

WALK-THROUGH SURVEY

Conduct a walk-through inspection of readily visible structural elements, taking notes that describe relevant information and adding sketches and photographs as appropriate. During the walk-through, observe readily accessible and easily visible structural elements and building service equipment. Generally, the walk-through survey will not include invasive probing, the removal of materials, or testing. The intent is to assess in a general manner the as-built conditions, the state of repair of the structural elements, and whether major building service equipments and nonstructural building elements are adequately restrained; this last item is especially important in earthquake-prone areas.

- General: With the initial screening factors and document review in mind, some questions to consider are:
  - □ Are there any undocumented and unpermitted structural modifications?

---

[10] Extracted from ASCE/SEI 31-03, Seismic Evaluation of Existing Buildings, pages 2–6. ASCE: American Society of Civil Engineers. SEI: Structural Engineers Institute.

- Are there visible weaknesses in structural members?
- Has settlement occurred or are there other foundation problems?
- Are there unusual structural features?
- Has all permanent service equipment, ducts, piping, and so forth, been properly (seismically) restrained? Note: Look especially at rooftop equipment such as water towers and large HVAC equipment.
- Concrete and rebar: Are there signs of any of the following:
  - Accident damage?
  - Alkali-aggregate reaction?
  - Cement soundness?
  - Contaminated water?
  - Cracking, scaling, spalling, and delamination?
  - Degradation?
  - Fatigue?
  - Sulfate attack[11]?
  - Freezing and thawing?
  - Mechanical damage?
  - Corrosion?
- Structural steel: Are there signs of any of the following?
  - Accident damage?
  - Corrosion?
  - Fatigue?
- Masonry: Are there signs of any of the following:
  - Accident damage?
  - Mortar and grout deterioration?
  - Loose bricks?
  - Surface deterioration?
  - Structural incompatibility?
  - Environmental attack?
  - Efflorescence and staining?
- Wood: Are there signs of any of the following:
  - Accident damage?
  - Decay/rot?
  - Excessive moisture?
  - Fire damage?
  - Insect damage?
- Foundations and substructures: Most building departments require that a soils and/or geological report be submitted prior to issuing a building permit for a large facility. Review this report and compare it to your observations. You may wish to contact the geotechnical firm that prepared the report to discuss its findings. If no report is available or problems such as settlement or sinkholes are observed, you may advise that a soils/geological investigation be conducted.

---

[11] Sulfate attack: With regards to concrete, a reaction between sulfate solutions and hardened cement paste. This reaction results from, for example, alkali magnesium and sodium sulfates in groundwater and seawater. The products of the reaction lead to expansion and cracking of the concrete. Sulfate is any salt or ester of sulfuric acid, such as sodium sulfate, Na_2SO_4, sodium hydrogen sulfate, or diethyl sulfate, $(C_2H_5)_2SO_4$. Adversely affects concrete and rebar. See Sulfate attack. Also spelled sulphate.

Your report should comment on footings, grade beams, foundation walls, waterproofing and under drains, insulation, and slabs-on-grade.

- ☐ Have buildings remained without settlement, alignment changes, and cracks in the foundation or building? Note: Look for shrinkage, settlement, heaving, and other subsoil failures that may have been caused by changes in the load-bearing capacity of soils due to shrinkage, erosion, or compaction/swelling; adjacent construction; structural or occupancy changes; and earthquakes.
- ☐ Are concrete floors sound (no serious cracks or arching)?
- ☐ Are drainage, drains, and sump pumps performing adequately?
- ☐ Is all roof drainage adequate to prevent basement/cellar moisture?
- ☐ Do exterior grade slopes carry water away from buildings?
- ☐ Are utility service lines and drains watertight?
- ☐ Have wall cracks, openings in construction joints, and so forth, been sealed?
- ☐ Are ventilation, vapor barriers, and humidification adequate?
- Surface material deterioration:
  - ☐ Are surface materials free of spalling, rebar corrosion, moisture penetration, and chemical reaction between cement and soil?
  - ☐ Are ferrous metals corrosion free? Note: Corrosion may be due to moisture or contact with acid-bearing soils.
  - ☐ Has wood decayed due to exposure, moisture, or insect infestation?
- Openings:
  - ☐ Do doors, windows, hatchways, and stairways function as designed?
  - ☐ Are utility penetrations sealed?
- Crawl space ventilation and maintenance:
  - ☐ Is air circulation adequate?
  - ☐ Are moisture barriers installed properly and are they adequate?
  - ☐ Are pest control and housekeeping satisfactory?
  - ☐ Are crawl spaces dry and clean?
  - ☐ Do air vents prevent animals from entering?

STRUCTURAL/SEISMIC DESIGN ASSESSMENT

Your walk-through inspection should determine if the building has adequate capacity to perform its intended use, if it is generally code compliant, and whether there are serious structural defects. If you are in a country prone to earthquake, you also need to consider seismicity, even though you are probably not a structural engineer (nor am I) and, therefore, cannot guarantee that all seismic design criteria have been met; your review is primarily a judgmental overview of the seismic capacity of the building(s); you need to ensure that your client understands this.

If structural drawings are provided, review them to discern the overall design approach and to identify the basic vertical and lateral load-resisting systems. Your review need not cover every detail of the building's construction, nor should you be expected to perform an in-depth analysis or provide detailed independent calculations. If the owner wants this then, as already stated, a structural engineer should be hired. The intent of a due diligence inspection is to assess the basic strengths and weaknesses of the building(s) from the standpoint of earthquake performance.

A structural engineer may have to be called upon to analyze the critical members and connections in the structural system to determine whether their resistant capacities for moment, torsion, axial, and shear forces meet current code requirements. If the structure has nonstructural members that may contribute to its structural resistance, these members should be included in the analysis.

- General:
  - Based on a review of the available documentation, results of an on-site inspection, performance criteria, and the anticipated loading, have the primary vertical and lateral force paths that transfer the forces to the foundation been identified? Notes: (1) Look especially for careless alterations that may have severed force paths. (2) Identify the code to which the building was built and compare with the current code. (3) Review available geotechnical and soils reports. Comment on performance of the foundations during life of building(s). (4) Identify other earthquake hazards at the site from specific geotechnical reports and local and USGS reports. These may come from local, state, and federal reports. Look especially at United Stated Geological Survey publications.
  - Are all bolts/rivets in place and do they appear sound?
  - Have the vertical, transverse, and longitudinal structural members and connections in the structural system been identified? Notes: (1) Critical members and connections are those whose failure would seriously reduce the structure's ability to resist applied forces. These should be identified. (2) Look especially at the building's lateral load system (roof/floor diaphragm, shear walls, etc.).
  - Are measures in place to reduce fire ignitions and spread and to improve fire suppression capacity following an earthquake. Note: Local fire departments may voluntarily assist in a building evaluation and may also suggest measures that should be implemented to reduce ignition and spread of fire.
  - Have ceilings, fixtures, pipes, and equipment been secured and do they comply with local codes and ordinances?
  - Have occupants been informed of earthquake risks and ways to reduce personal risk?
  - Are water and gas shut-off valves in place? Note: Maintenance personnel should know the location of all shut-off valves.
  - Are the buildings in danger of earthquake-induced landslides?
- Material properties:
  - Do existing drawings contain information about the material properties that were used in the design? Note: If existing drawings are not available, estimate the strengths of materials based upon design criteria and the types of structural materials that were used when the building was constructed.
  - Is further testing required? Note: Usually testing in not required for a due diligence inspection unless there are signs of distortions and potential failure.
- Façades, parapets, and decorative features:
  - Is there a record of periodic façade inspections that include façades, parapets, and decorative features that are fixed to the building's exterior? Note: Most major cities, for example, New York City and Chicago, have

façade inspection laws that require such inspections; these inspection criteria are readily available.

❑ Are parapets and decorative features fixed and safe from falling onto pedestrians? Note: Old terra cotta ornamentation is often both brittle and not well fastened.

- Earthquake Loss Assessment: Per ASTM E 2026, a Level 1 investigation requires the highest general experience in professional practice and evaluation. In other words, a usual due diligence inspection would not include such an assessment unless the owner or proposed owner requests. Then a professional engineer would have to be included on the inspection team. An earthquake loss assessment would be a separate section in the inspection report. ASTM not required. There is a previous footnote.

A Level 1 Investigation is defined in the ASTM document E 2026 entitled "Standard Guide for the Estimation of Building Damageability in Earthquakes." Generally, it is an estimate of the direct physical damage to the building(s) due to an earthquake's strong ground shaking; other site seismic hazards such as tsunami, liquefaction, and surface fault rupture are considered in the damage estimates. This physical damage estimate is expressed as a percentage of replacement cost of a building's structural and nonstructural elements only; building contents are not included. Also not included are social losses and indirect losses such as business interruption or fire following an earthquake. The loss estimates are usually based on motion–damage relationship data based on past studies by others, proprietary methodology, and data that has been gathered during an inspection. The code to which a structure was designed, unusual deficiencies or strengths in the seismic force resisting system, general state of repair, and vulnerability of nonstructural elements and building equipment also need to be factored into the earthquake loss assessment. For a given intensity level and class of building, there will be a range of damage actually experienced due to variations in the earthquake motion and in the structural characteristics of the individual buildings in each building class. Generally estimated mean damage level and the 90th percentile damage level are determined.

ASTM Standard E 2026-99 "Standard Guide for the Estimation of Building Damageability in Earthquakes," defines the mean loss as the Scenario Expected Loss (SEL) and the 90th percentile loss as the Scenario Upper Loss (SUL). The 90th percentile estimate of loss means that, on average, only 10% of similar buildings will experience greater damage for the same ground motion. Clearly, there is a great deal of uncertainty in estimating both the earthquake ground-shaking intensity and the building damage. Therefore, there is also a great deal of uncertainty in the loss estimate.

BUILDING EXTERIOR[12]

The building envelope gives clues as to the building's age and life expectancy, how well it has been maintained and potential durability, whether it is code compliant, whether life safety and disabled access have been considered when it was built and during its life

---

[12] The primary reference for this section is: SEI/ASCE 30-00, Guidelines for Condition Assessment of the Building Envelope.

and any number of special purposes based on the specific building, and its current or proposed occupancy or function.

The building envelope comprises a set of components that have been assembled to separate or selectively filter the outside environment from the building's interior. The components are usually made up of dissimilar materials having dissimilar properties, which have been assembled to work as a system. On a broader scale, the building envelope must interface and interact with other systems such as the structural, HVAC, and lighting systems. To properly assess the building envelope, you must be aware of these individual materials, as well as of the total system.

Recent events such as Hurricanes Katrina and Sandy, the Northridge (California) Earthquake, and typhoons and earthquakes in other countries have wrought havoc across wide areas. Even where the 2006 International Building Code (IBC) had been adopted and enforced, structures crumpled in their paths. As a result, contractors, engineers, architects, and manufacturers of construction products have gone back to their drawing boards for what seems to be an endless battle against nature.

Moisture and water penetration play a large role in the overall design of a building envelope and so too do vapor and air migration. The primary components of interest are exterior walls, exterior windows, exterior doors and frames, entrances, chimneys, and exhaust stacks. Inspectors generally have little say in the design of the building's exterior or its resistance to shear forces, but they do play a major role in finding and identifying problem areas in existing buildings. When it comes to exterior walls, even seemingly small deficiencies may have an enormous effect on the letting in and draining out of moisture and water.

EXTERIOR WALL ASSEMBLY

Generally, wall systems consist of the following basic elements, or layers: exterior cladding (natural or synthetic), drainage plane(s), air barrier system(s), vapor retarder(s), insulating element(s), and structural elements. Wall systems are either load bearing or non–load bearing. A load-bearing wall system supports a vertical load in addition to its own weight, whereas a non–load-bearing system supports only its own weight. Both systems have similar components.

Either system may serve as a shear wall, in which case it is required to resist in-plane lateral forces and transmit them into the foundation or ground as a part of the main lateral force resisting system (MLFRS) for the structure as a whole. All wall systems must be capable of resisting out-of-plane lateral forces between their supports and transferring them into the foundation or ground or to the MLFRS. Some non-bearing wall systems, for example, veneer walls, can only transmit lateral forces to the MLFRS.

When inspecting a wall system, it is important that you understand the manner in which the wall was intended to resist water penetration and thus anticipate problems.

- Barrier wall systems are systems that allow moisture to be absorbed within the material, but not go completely through it. They can be either load bearing or non–load bearing. Examples of barrier wall systems include
  - ▫ Unreinforced and reinforced concrete and precast concrete:
    - – Are structures sound? Notes: (1) Record any repairs that may be required. Look especially at exterior concrete spandrel beams,

concrete fasciae, and for exposed rebars. (2) Make sure that balcony handrails are properly anchored, checking especially for rust where the handrail and concrete meet.

- Are embedded steel plates or shapes for connection purposes solid?
- Is insulation present and is it dry?
- Has a vapor barrier been installed and is it functioning as intended?

☐ Unreinforced and reinforced masonry, unreinforced and reinforced cavity walls, reinforced masonry cavity walls, and masonry veneer with wood or steel stud back-up:

- Are bricks firmly planted? Notes: (1) Check to see if wythes[13] have moved, cracked, or become lose. (2) Identify individual bricks that have to be replaced. (3) Make sure that the bricks are anchored to the underlying structure.
- Are parapet walls sound? Check copings[14] to make sure they are keeping water from entering the parapet wall, and flashings to make sure they are keeping water from entering under the roof.
- Are parapet wall heights in compliance with the latest code? Note: Requirement for parapet walls vary based on such factors as a building's classification and type of construction; therefore, you must review the applicable code.
- Are expansion or control joints functioning as intended? Note: Look for bulging that may be caused because of insufficient expansion or control joints.
- Are flashing, weeps, and scuppers in satisfactory condition and functioning as intended? Check to ensure that galvanic action cannot occur.
- Are masonry walls free of efflorescence and staining?
- Is there surface deterioration? Look especially at the higher levels of the building.

☐ Drainage wall systems: Drainage wall systems assume that moisture enters the system. This moisture is diverted back to the exterior through a system of cavities, flashings, and weep-holes. They can be either load bearing or non–load bearing:

- Are weep holes clear of debris?
- Are masonry walls free of efflorescence and staining? Note: Efflorescence and staining are signs that water may be (probably is) trapped within the wall system or assembly.

• Surface-sealed systems: Surface-sealed systems are intended to be impermeable to moisture and air at the exterior face. Several types of curtain wall systems, for example, aluminum frame curtain walls and interlocking metal panel walls, are surface-sealed systems:

☐ Has water penetrated the system? Have the supports rusted or rotted?

---

[13] Wythe, or withe: A vertical section of masonry that is one unit thick, that is, it is constructed from a single line of masonry units.

[14] Coping: The top layer or course of a masonry wall. Copings usually have a slanting or arched upper surface to shed water.

- ☐ Are panel gaskets/sealants deteriorating?
- ☐ Are there signs of leakage around the windows?
- ☐ Are fasteners on metal building tight?
- ☐ Are metal panels rusted/corroded? Is the damage due to impact?
- • Rain screen/pressure equalization wall systems: These systems are used primarily in curtain walls. Some aspects of the theories used with these systems are also used in masonry cavity wall and veneer wall systems. Generally, these systems have two exterior faces. The outer system is intended to keep water out, but recognizes that some water will enter by gravity, kinetic energy, surface tension, capillary action air currents, and/or pressure difference. The interior wall controls air infiltration. The space between the "rain screen" and the inner wall is ventilated to the outside to allow the air pressure within the space to equalize with the exterior air pressure. Walls are generally made of glass, aluminum extrusions, or concrete panels.
  - ☐ Have these different materials been functioning as intended? Note: Check to see if water has penetrated the system. Look at the supports for rust and rot.
  - ☐ Are panel gaskets/sealants lively? Note: If gaskets and sealants are dry and brittle, they may have to be replaced, which could be very expensive.
  - ☐ Are there windows tight? Note: Look at interior walls and ceilings for water stains.

FENESTRATION

A fenestration system is a non–load-bearing element that is installed within a wall. Fenestration systems include all types of windows, louvers, doors, and entrances. Materials and configurations vary greatly, but there are some basics you should follow when inspecting windows for a due diligence/capital planning program inspection:

- • Windows—General:
  - ☐ Are window frames square? Note: Normally widows should not be more than ¼-inch out of square. If they are not, replacement may be in order.
  - ☐ Is the trim sound? Note: You should also note that the trim matches the esthetics of the building or if it has historic worth.
  - ☐ Are there signs of water damage?
  - ☐ Have windows been replaced with low-quality or inappropriate windows? Note: There is a battle raging between "sustainability people" and "historical preservation" people with the historical preservation people often requiring single-pane windows to match the original while owners may be more interested in conserving energy and, thus, money. If this applies it may warrant a note in your inspection report.
  - ☐ Do perimeter joints prevent water leakage? Note: Water leakage can come from poor glazing, inadequate perimeter sealant, and adjacent walls; look especially at the corners.
  - ☐ Do window systems interface properly with their structures? Note: Look at headers to ensure that they create water barriers. Note: Excessive air and water penetration may be the result of the way windows have been integrated into surrounding walls. Check interior walls for water stains. If necessary, an invasive inspection may be necessary.

- Have all codes been met: Note: Windows are affected by numerous codes, among them: energy (U-factors, solar heat gain, air infiltration), structural (wind pressure), safety, and ADA (egress requirements, safety glazing, EPA regarding renovation repair and painting).

Many types of fixed and operable windows use the rain screen and pressure equalization principle to resist moisture and air infiltration, in generally the same fashion that is used in aluminum frame curtain wall systems. The application of pressure equalization design to fenestration products is not material dependent. Because of the increased application of this design principle on high-rise buildings, the majority of these frames tend to be fabricated from aluminum extrusions:

- Extruded aluminum, PVC, or wood frames and sashes:
  - Are frames airtight?
  - If painted, what is the condition of the paint?
  - Is there rot?
  - What is the condition of weather-stripping?
  - Are screens intact?
- Single pane or insulated glass panels:
  - Are panes firmly supported on all sides or subjected to unusual load conditions? Note: Check drawings to ensure construction is per design.
  - Are setting blocks[15] spaced and sized properly?
  - Are glazing seals, air seals, sealants, and backer rods in satisfactory condition?
  - Has double-pane windows free of inside moisture?
  - Are flashings and weeps functioning as intended?

HARDWARE

Hardware receives an incredible amount of use and abuse, especially exit/entrance doors and doors in special use facilities such as schools. In large facilities, a logical keying system is crucial for security, fire safety, and maintenance/janitorial operations; lock manufacturers can help you with this. Your inspection should include a careful look at the hardware throughout the facility:

- Are locks, latches, hinges, and so forth, of high quality?
- Are locks, latches, hinges, and so forth, well maintained?
- Is a logical keying system in place?
- Are closure devices operable?
- Is panic hardware in place and operable?
- Can all exterior doors be opened in the event of an emergency? Note: It is not uncommon for a building's facilities staff to chain little used doors, for example, basement doors of schools through which truants and hooligans sneak. This is not permitted.

---

[15] Setting block: A small synthetic block of material used to support the glass at its base bearing surface. Setting blocks are also installed inside the window sash, between the sash material and insulated glass unit, to protect the insulated glass from the impact of opening and closing the sash and to provide same stress relief as vinyl windows expand and contract over varying temperature conditions. The setting blocks keep the glass from cracking and serve to protect the integrity of the glazing bead.

MISCELLANEOUS

Inspectors should also consider the following and make comments as appropriate:
- Acoustical resistance, especially for schools, libraries, laboratories, and buildings located near airports, highways, and railroads.
- Any unusual esthetic treatments or such things as embedded logos that a new owner may want to remove.
- The durability of materials.
- Fire safety measures, especially at the intersections of walls and floors.
- Serviceability/maintainability and water and air tightness. Consider control joints, expansion joints, and isolation joints.
- Physical security. Note: Note especially areas into which a victim could be dragged and assaulted/raped, for example, unlocked closets, beneath stairs.[16]
- Wildlife, insect, and organism resistance.
- Problems and potential problems associated with air infiltration.
- Deterioration due to inadequate control of condensation developed.
- If applicable, the historic significance of the facilities.

ROOFING

The life of a roof depends on factors such as weather conditions, design and installation, the quality of the materials, and maintenance. Slate, copper, and clay/concrete roofs have the longest life expectancy (some of Harvard University's slate roofs are original). Roofs made of asphalt shingles last for about 20 years, roofs made of fiber cement shingles for about 25 years, and roofs made of wood shakes for about 30 years. Some common membranes such as atactic poly propylene (APP), styrene butadiene systems (SBS), and thermo plastic olefin (TPO) have various life spans ranging around 20 years.

Since roofs are generally long lived they deteriorate incrementally, a bit at a time. So the question you will be asked to answer is not simply whether it is satisfactory or unsatisfactory. You will probably be asked for your opinion on how long it will be before the roof will have to be replaced and what can be done to extend its life. These questions are not necessary in the purview of an inspector, but most likely unavoidable. Unless you feel confident about your ability to answer them, you may wish to suggest hiring a roofing expert. You may also consider recommending moisture meter readings[17] or infrared readings if the roof is old or in any way

---

[16] In the case of *Jennifer C. v. Los Angeles Unified School District*, B2055903, the California Court of Appeal ruled that the maintenance of a "hidden location" where a 14-year-old could be victimized satisfied the foreseeability factor of the duty analysis, even if the school was not aware of any prior sexual assaults or illegal activities taking place in the alcove beneath a flight of stairs on a middle school campus.

[17] Moisture meters detect moisture just below the existing roofing membrane and also deeper within the existing roofing system. Moisture meters are usually nondestructive, that is, they do not cause any surface damage to the membrane. Pin-type meters can sometimes be used to qualify results once problem areas have been detected and opened up, but they are not usually used for carrying out an overall moisture survey. Moisture meters have adjustable sensitivity that can alter the depth of penetration of the signal. Therefore, they can be used to suit the particular roofing system under test.

For built-up and single-ply roofing, moisture meters are designed to detect elevated moisture within the insulation and thickness of the roof and also assist in tracing leaks back to source. They can also identify areas of inter-ply moisture.

problematic. Meanwhile, take lots of pictures and keep good notes that identify observed problems and potential problems.

WATER TIGHTNESS

I'll assume you will get the formatting correct.

- Initial assessment: Walk all over the roof always looking for problems and potential problems. Consider the following and comment as appropriately. Take photos to make your point.
  - Evidence of leaks on the undersurface.
  - Surface weathering.
  - Faulty or inappropriate materials. Note: Look especially at flashings where shortcuts may have been taken.
  - Faulty design. Again, look at flashings, scuppers and drains, material overlaps, and so forth.
  - Poor construction. Note: Check for fastening failure.
  - Standing water and weather damage.
  - Accumulations of leaves and trash.
  - Parapet walls. Note: Look for cracks, land loose, and missing bricks and copings.
  - Antennas, signs, solar panels, and other attachments. Note: Look for poor installations.

CONSTRUCTION

Identify the roofing system and comment as necessary:

- Felt or bitumen built-up:
  - Adhesion.
  - Bare areas.
  - Cracks, holes, tears.
  - Alligatoring.[18]
  - Blisters, wrinkles.
  - Fish mouths.[19]
  - Ballast. Note: Ballast should be evenly spread and such that it cannot be blown off the roof by heavy winds.
- Single-ply (Thermosetting, thermoplastic, composites): Are there signs of any of the following:
  - Blisters, wrinkles?
  - Holes, tears?
  - Protective coatings?

---

[18] Alligatoring: A network of fine cracks on a surface with a definite pattern as indicated by the name. For example, paved highways, roofs, and painted surfaces frequently alligator. Weather aging often causes the effect.

[19] Fishmouth: With regard to roofing, a half-cylindrical or half-conical shaped opening or void in a lapped edge or seam, usually caused by wrinkling or shifting of ply sheets during installation; in shingles, a half-conical opening formed at a cut edge. Also referred to as an edge wrinkle.

- Metal roofing: Are there signs of any of the following:
  - ☐ Corrosion? Note: Estimate the percentage of corrosion.
  - ☐ Seams?
  - ☐ Holes?
  - ☐ Protective coating?
  - ☐ Cracks or breaks?
  - ☐ Insulation? Note: record the type of insulation and how it is attached.
- Shingles and tiles (Metals, clay, mission, concrete, other): Are there signs of any of the following:
  - ☐ Disintegration?
  - ☐ Missing, broken, or cracked shingles? Note: Try to estimate the number of shingles or square feet that need to be replaced.
  - ☐ Underlayment? Note: Seeing the underlayment is usually not a good sign.
  - ☐ Ineffective fasteners?
- Wood shingles: Wood shingles are out of favor because of their fire hazard. Local codes may require replacement. Are there signs of any of the following?
  - ☐ Cracked?
  - ☐ Curled and moldy?
  - ☐ Missing shingles?

OTHER ROOFING DETAILS

- Drainage: Slope and rigidity are the critical factors in preventing accidental ponding of water and ensuring positive drainage of the roof. For low-slope roofs, the normally recommended live-load deflection limits are from 1/180 to 1/360 of the span, but even these deflections may result in ponding.
  - ☐ Does alignment and pitch allow for proper drainage? Note: Is the roof's slope to drainage at least ¼ inch per foot?
  - ☐ Is there standing water or are there signs of standing water?
  - ☐ Has trash, leaves, and so forth, accumulated? Note: Check especially around drains, as clogged drains frequently impound water.
  - ☐ Are gutters and downspouts secure?
  - ☐ Are drains, gutters, and downspouts clogged?
  - ☐ Check for corrosion.
- Base/wall flashing:
  - ☐ Are flashing and nailing strips secure?
  - ☐ Is base flashing a minimum of 8 inches above the highest waterline?
  - ☐ Is flashing continuously supported?
  - ☐ Are nails compatible with flashing material (to avoid galvanic action)?
  - ☐ Are gravel stops elevated above the general roof level to prevent water from penetrating the roof's edge?
  - ☐ Is differential movement accommodated at gravel stops (it should be)?
  - ☐ Is differential movement between base and cap flashings possible (it should be)?
  - ☐ Has through-the-wall flashing been installed to prevent water from filtering through the core of masonry walls?

- Coping:
  - ☐ Does coping fully cover all parapet walls?
  - ☐ Is there sufficient overlap to prevent water penetration?
- Penetrations:
  - ☐ The best flashing detail at larger roof penetrations follows the general rule of attaching the base flashing to the structural deck and counterflashing to the penetrating element. Has this practice been followed?
  - ☐ Are small pipes, vents, and similar roof penetrations properly sealed? Notes: (1) Usually, small pipes, vents, and similar roof penetrations do not have separate support for base flashing. Instead, the flanges of metal or plastic sleeves are attached directly to the roofing membrane, and the flanged sheet metal support carries the base flashing around it. (2) Counterflashing attached to the pipe shields the base flashing. The flanged sheet metal support, or pan, is often filled with soft mastic. This is difficult to inspect. (3) Check especially, to ensure that leaking has not previously occurred.
  - ☐ Have penetrations been installed according to plans?
- Exterior terraces and balconies: Generally, the same principles apply to terraces and balconies as to roofs, except many allow people to gather on them. Your inspection needs to consider the past and future effects of this additional wear and tear. Here you are correct, it should be effects.
  - ☐ Is the membrane that is usually a coating that is applied to a concrete substrate, able to serve as the barrier to moisture? Is it capable of withstanding traffic and direct exposure to weather?
  - ☐ Does the terrace/balcony drain properly from area drains or scuppers?
- Expansion joints:
  - ☐ Are expansion joints no more than 300 feet apart?
  - ☐ Do they permit movement across the joint and shear movement parallel to the joint?
- Insulation: Generally, insulation systems can be divided into four categories (identify and describe condition as best you can): (1) rigid insulation, prefabricated into boards that are applied directly to the deck surface; (2) dual-purpose structural deck and insulation planks; (3) poured-in-place insulating concrete fills; and (4) sprayed-in-place plastic foam.
  - ☐ Has the insulation maintained its dimensional stability under thermal changes and moisture absorption?
  - ☐ Is compressive strength sufficient to withstand traffic loads?
  - ☐ Does insulation show signs of disintegration?
  - ☐ Has moisture gotten into the insulation?
- Roof hatches, smoke hatches, skylights:
  - ☐ See Base/wall flashing, above.
  - ☐ Are all glazing and coverings on hatches in good condition?
- Ventilation: Properly designed sub-roof ventilation is the best technique for preventing water-vapor infiltration into a membrane roof. Inspect to ensure that moisture is being diffused from under the membrane roof.
- Walking surfaces:
  - ☐ Are walking surfaces placed logically?
  - ☐ Is there evidence of people walking when pavers do not exist?
  - ☐ Do pavers, and so forth, block drainage?

- Railings:
  - ☐ Are railings secure?
  - ☐ Are railings painted?
- Heat island reduction[20]: This is an issue that is moving to the forefront as owners seek LEED certification.
  - ☐ Have measures been taken to reduce heat islands (thermal gradient differences between developed and undeveloped areas) to minimize impact on microclimate and human and wildlife habitat?
  - ☐ Does the roof have ENERGY STAR® compliant, high-reflectance and high-emissivity roofing material that has a minimum emissivity of 0.9 when tested in accordance with ASTM 408? What percent of the roof is covered with this material?
  - ☐ Does the building have a "green" (vegetated) roof? What percent of the roof is "green"?
- Warrantees: Review the warrantee and report on the following:
  - ☐ Whether the warrantee is transferable to the new owner?
  - ☐ Whether the warranty requires periodic (~3 years) inspections?
  - ☐ The warranty's duration (10/15/20 years).
  - ☐ Does the warranty specify "Materials and Workmanship guarantee?" Note: Check to see if any contractor can make repairs should materials and workmanship be found defective.
  - ☐ Whether the warranty is backed up by the roofing materials manufacturer?

BUILDING INTERIOR

Just as the exterior gives clues about the building, so too does the interior. Indeed, users are more likely to notice the condition of lobbies, rest rooms, reception areas, hallways, offices, eating areas, and other public access areas that they are the grounds, unless, of course the grounds are especially beautiful or rundown.

You are likely to run across people who work in the building. Don't hesitate to ask these people questions about the building and its operations.

ENTRANCES

This is the first welcome people get when they arrive. Beyond appearance, well-designed and maintained entryway systems "reduce exposure of building occupants and maintenance personnel to potentially hazardous chemical, biological, and particle contaminants that adversely impact air quality, health, building finishes, building systems, and environment."[21] It is not unusual for custodians to say that 80% of all dust and dirt in buildings is tracked in on shoes.

- Are doors in good order?
- Is lettering accurate and neat? Note: Check for chipped and cracked lettering.

---

[20] References: U.S. EPA. ENERGY STAR® Roof Compliant High-Reflectance, and High Emissivity Roofing. Lawrence Berkeley National Laboratory Heat Island Group – Cool Roofs. http://eetd.lbl.gov/HeatIsland/CoolRoofs/
[21] LEED for Existing Buildings v2.0 Reference guide, page 439.

- Is the hardware, for example, door knobs and panic hardware, in good condition? Is it of high quality?
- Do the doors swing easily?
- Do revolving doors turn easily? Are there signs of them being out of alignment?
- Is the entrance welcoming?
- Are entryways equipped with grills, grates, mats, and so forth?
- Are thresholds maintained and clean?
- Are entryways well lit?

PUBLIC REST ROOMS, SHOWER ROOMS, KITCHENS, AND FIXTURES

- Rest rooms:
  - Are floor and wall tiles in place and clean?
  - Is the grout between tiles sound and clean?
  - Are there cracked or missing tiles?
  - Are there soap dispensers and are they operable?
  - Are there towel dispensers and/or air dryers and are they operable? Note: Identify to the extent possible the types of hand dryers, and if paper towels are used the supplier of the towels.
  - Do the toilets and urinals function properly? Note: Identify to the extent possible the types of fixtures that are installed. Note especially if old, high-water use fixtures are still in place.
  - Do sinks and faucets function properly?
  - Do the rest rooms have unpleasant odors? Note: Look to see if high-powered deodorizers are installed to hide odors.
- Bicycle facilities and shower rooms: In an effort to reduce pollution and the adverse effects of land development from automobile use, some buildings are now providing bicycle storage and changing rooms that include showers; the city of Los Angeles is going to great lengths to encourage bicycle riders to commute. LEED offers points for such facilities. Now I'm getting confused, but I believe automobile use affects land development, the effect being urban sprawl. I'll have to bow to you on these matters.
  - Are bicycle racks sufficient to support bicycles by their frames, allowing for the frame and one or both wheels to be secured?
  - Are racks constructed of durable, sturdy materials?
  - Is there sufficient spacing between individual bicycle parking spaces to prevent damage?
  - Are lockers and changing spaces adequate? Note: Note their state of repair and cleanliness.
  - Are showers satisfactory? Notes: (1) LEED requires one shower per every eight bicycle racks. (2) Check to ensure that floor and wall tiles are in place and clean and free of fungus. Make a note of cracked and missing tiles. (3) Make sure that the grout between tiles is clean and free of fungus. (4) Make sure that shower heads function properly.
  - Do the locker and shower rooms have unpleasant odors?
- Kitchens: Inadequate and unclean kitchens attract vectors of all types. They should be large enough to support the staff, clean, and in good condition.

Many of the same questions that apply to restrooms and locker/shower rooms apply to kitchens.

- Fixtures: See Plumbing fixtures.
- Ceilings: Ceilings are usually exposed structural systems, directly applied ceilings, or suspended systems.
  - ☐ Is overall appearance satisfactory and do the ceilings meet expected standards? Notes: (1) Are all units in place? (2) If the ceiling is expected to have acoustic quality it should not be painted and, thus, reflect instead of absorb sound.
  - ☐ Are ceilings level, properly aligned and attached? Notes: (1) Check for settlement and sagging. (2) Check also for evidence of moisture and water damage.
  - ☐ Are lighting fixtures secure?
  - ☐ Do louvers permit proper air/heat distribution?
  - ☐ Are sprinklers/fire protection in place and free of obstructions?

FLOOR COVERINGS

You should familiarize yourself with the many types of flooring: carpet, composition, concrete, resilient, ceramic tile, masonry, terrazzo, wood, metal, and others such as raised floors and so forth.

- Is the overall appearance satisfactory and do the floors meet expected standards? Note: Look for visible settlement, evidence of moisture, irregular surfaces, tripping hazards, and accessibility hazards.
- Carpets (tufted, tile): Are there signs of any of the following:
  - ☐ Age?
  - ☐ Wear?
  - ☐ Stains?
  - ☐ Discoloration?
  - ☐ Holes/tears?
  - ☐ Seam conditions?
- Monolithic toppings (concrete, granolithic, terrazzo, magnesite): Are there signs of any of the following:
  - ☐ Cracks?
  - ☐ Porosity?
  - ☐ Open joints?
  - ☐ Missing or popping sealing?
  - ☐ Poor janitorial practices?
- Resilient flooring (asphalt tile, cork tile, linoleum, rubber, vinyl): Are there signs of any of the following:
  - ☐ Broken and loose tiles?
  - ☐ Shrinkage?
  - ☐ Lifting, cupping?
  - ☐ Cuts, holes?
  - ☐ Fading?
  - ☐ Porosity?
  - ☐ Poor janitorial practices?

- Wood (plank, strips, block, and parquet): Are there signs of any of the following:
  - Shrinkage?
  - Cupping, warping?
  - Excessive wear?
  - Unevenness?
  - Decay?
  - Sealing?
  - Stains that cannot be easily removed?
- Other (raised floors, etc.): Ensure that the floors have been installed properly. Check to ensure that proper fire blocking has been installed and that trash has not collected.

INTERIOR WALLS AND PARTITIONS

This section includes interior walls; wall coverings and finishes; interior doors; windows and frames; hardware; special openings such as access panels, shutters, etc. Consider:

- Partitions, framing, and movable walls: Inspect and report on the following:
  - Strength and stability
  - Physical strength
  - Acoustic quality
  - Signs of moisture, stains
  - Maintainability
  - Adaptability
  - Code compliance
  - Abuse, vandalism
- Interior doors, windows, and frames: Inspect and report on the following:
  - Frame condition. Note: If there is apparent rot around door and window frames and sills, probe deeper for termite damage and water leakage.
  - Frame anchoring
  - Door surfaces
  - Glazing
  - Seals, air, and water tightness
  - Shading devices
- Wall coverings and finishes: Inspect and report on the following:
  - Cracks
  - Joint openings
  - Peeling, flaking
  - Rips, tears
  - Looseness
  - Signs of moisture, stains
  - Missing segments
  - Abuse, vandalism
- Hardware: Inspect and report on the following:
  - Overall condition
  - Maintainability
  - Operation

- ◻ Keying system, locksets
- ◻ Closure devices
- ◻ Panic hardware
- ◻ Bolt locks. Note: Most codes do not allow manually operated flush bolts or surface bolts wherever emergency exiting may be required.

MISCELLANEOUS

Inspect and report on the following:

- • Specialty items:
  - ◻ Projection screens.
  - ◻ Signage. Notes: (1) Check especially for exit signs, including floor-level signs for exiting in the event of fire. (2) If people have to move from place to place as in a hospital, check to see if the signage is logical.
  - ◻ Telephone enclosures. Note: With the advent of cell phones, these are almost extinct.
  - ◻ Blinds.
  - ◻ Window coverings to prevent glare.
  - ◻ Waste handling? Notes: (1) Has space been set aside for recycling for, at a minimum, paper, glass, plastics, cardboard, and metals?(2) Check for odors and signs of vermin.

RECYCLING GUIDELINES

COMMERCIAL BUILDING AREA (SF)	MINIMUM RECYCLING AREA (SF)
0–5,000	82
5,001–15,000	125
15,001–50,000	175
50,001–100,000	225
100,001–200,000	275
200,001 and more	500

- • Support spaces: It is generally wise to have a guide from the facilities maintenance department to accompany you when entering Mechanical, Electrical, and Elevator Rooms (For this section your purpose is not to inspect the equipment or its operation, but to gain an impression of the area itself).
  - ◻ Are doors locked?
  - ◻ Are areas painted, clean, and orderly?
  - ◻ Are storage areas in compliance with fire codes?
  - ◻ Are there signs warning of low overhead hazards?
  - ◻ Are rooms well lit?
  - ◻ Are there unpleasant odors and signs of vermin?
  - ◻ Is there standing water or oil on the floor?
  - ◻ Are temperatures reasonable?
  - ◻ Are emergency telephones available?
  - ◻ Where required by code, are fire extinguishers available?

- Residential areas: If people are living in the facility here are some items to check:
  - ☐ Has a Radon check been conducted?
  - ☐ Is there a smoke detector in every sleeping area?
  - ☐ Are all handrails secure?
  - ☐ Do all fire extinguisher gauges show green?
  - ☐ Are fence hinges and locks around pools in good shape? Are latches on automatically closing gates at least 54 inches above the ground?

See also: Electrical rough-in; Mechanical rough-in; Plumbing rough-in and Final inspections

LIMITED DISABLED ACCESS REVIEW

Generally, a due diligence inspection does not include a thorough, 100% ADA review; there are people, usually architects, who are experts in ADA. You should, however, make note of obvious violations and alert the owner and potential owner about them and include these in your cost estimates.

The ADA was written into law in 1990 and became effective on January 26, 1992. The purpose of ADA is to allow people with disabilities to integrate into and contribute to society to the maximum extent possible. You should keep this in mind as you conduct your inspections. Sometimes ADA is more restrictive than local codes and sometimes not; however, the more restrictive portions of either standard should be implemented. Accordingly, all tenant spaces of building and parking areas should be designed to be accessible. All future renovation work will have to meet the requirements of the more restrictive standard, that is, ADA or local ordinances.

Title I of the ADA states that it is the responsibility of the employer to make a disabled employee's workplace accessible. Title III of these federal civil rights regulations requires that public buildings provide access to the disabled. The common areas of multi-tenanted buildings are typically defined as public accommodation.

Therefore, due diligence/capital planning inspectors should make a good faith effort to identify any barriers to the disabled and other issues that are now commonly known to the architectural and facilities engineering profession. Generally, when non-compliant building spaces or components exist, but making them compliant is "readily achievable,[22]" you should include this in your report.

PARKING AND EXTERIOR ACCESSIBLE ROUTES

Here are some typical parking requirements, but as you prepare your inspection checklist check the local ordinances too, as codes vary from community-to-community.

- Common parking requirements:
  - ☐ Do entrances to parking areas provide vertical clearances of ≥ 9 feet 6 inches?
  - ☐ Are there sufficient reserved accessible parking spaces? Note: See table below.

---

[22] "Readily achievable" is usually interpreted to mean that the work is "easily accomplished without much difficulty or expense." Needless to say "without much difficulty or expense" is open to interpretation.

MINIMUM NUMBER OF ACCESSIBLE PARKING SPACES[†]

ADA STANDARDS FOR ACCESSIBLE DESIGN 4.1.2 (5)			
TOTAL PARKING SPACES IN LOT OR GARAGE	**REQUIRED NUMBER OF VAN ACCESSIBLE**		
	ACCESSIBLE PARKING SPACES(60" & 96" AISLES) COLUMN A	**ACCESSIBLE PARKING SPACES MIN 96" AISLE**	**ACCESSIBLE PARKING SPACES MIN 60" AISLE**
1–25	1	1	0
26–50	2	1	1
51–75	3	1	2
76–100	4	1	3
101–150	5	1	4
151–200	6	1	5
201–300	7	1	6
301–400	8	1	7
401–500	9	2	7
501–1,000	2% of total parking provided in each lot	1/8 of Column A*	7/8 of Column A**
≥1,000	20 plus 1 for each 100 over 1,000	1/8 of Column A*	7/8 of Column A**

* One out of every 8 accessible spaces.
** 7 out of every 8 accessible parking spaces.
[†] Taken from "ADA Design Guide," U.S. Department of Justice, Civil Rights Division, *Disability Rights Section*.

Also:

1. For a shopping center or facility having at least five separate retail stores and at least 20 but not more than 500 public off-street parking spaces, a minimum of 5% of the total number of parking spaces or 10 spaces, whichever is less.
2. For an outpatient medical unit or facility, a minimum of 10% of the total number of parking spaces serving each such unit or facility.
3. For a facility that specializes in treatment or services for persons with mobility impairments, a minimum of 20% of the total number of parking spaces.

   □ Are the accessible parking spaces located closest to the accessible routes and accessible building entrances?
   □ Does the accessible space measure 96-inches wide with an adjoining access aisle of 96-inches wide?
   □ Are accessible spaces identified with permanently installed signs that are 5 to 7 feet above grade? Note: These signs should include the international symbol of access.
   □ Are all slopes along the accessible route less than 1:20?
   □ Do the access aisles have a cross slope ≤1:50? Also, surfaces should be firm, stable, nonslip.

▫ Do the access aisles connect to accessible pedestrian routes that have a clear and unobstructed width of ≥36 inches? Note: Where the entire level of the accessible route is 36 inches wide, 60 inches x 60 inches passing spaces should be provided at least every 200 feet.

▫ Are there passenger pick up and drop off zones? Note: If so, there should be at least one accessible passenger loading zone that measures ≥20-feet long and has a ≥5-feet wide access aisle that is parallel to the vehicle pull-up space and is at the same level as the roadway. (The access aisle is the beginning of the accessible route.)

▫ Do curbs on accessible routes have curb cuts or curb ramps at ≤1:12 slope? Note: Curb cuts/ramps should be ≥36 inches wide, exclusive of flared sides.

▫ Are there public transportation stops on site? Note: If so, accessible routes should be provided to the building from the stop. Surfaces should be firm, stable, and slip resistant, with no cracks.

▫ Does at least one accessible route connect accessible buildings, accessible elements, and spaces that are located at the same site?

▫ Does the accessible route provide for a clean, unobstructed width of at least 36 inches? Notes: (1) If any object is protruding into the accessible route, it should be detectable by a person who has a visual disability and is using a cane; an object must be within 27 inches from the ground to be detected by a cane. (2) Objects that are hanging or mounted overhead must be higher than 80 inches to provide clear headroom. (3) There should not be any grates along the accessible route.

• Entrance/exterior doors: By their very nature, entrance doors are an obstruction for people who are mobile impaired; if not in compliance with ADA, a door is a lawsuit waiting to happen. A building that has one entrance only up a flight of stairs, with no wheelchair ramp, violates the law. Exterior entryways must also connect via a wheelchair-accessible route to public transportation stops, to accessible parking and passenger-loading areas, and to public streets and sidewalks. Exterior doors must link to all wheelchair-accessible areas inside the building. A service entrance cannot be the only accessible entrance into a building, unless it is the only entryway into the building, for example, into a garage. You should be conscious of the disabled when conducting a due diligence/capital projects planning inspection.

• Not every building needs upgrades to comply with the ADA; the law is not retroactive. The rules went into effect on January 26, 1992; therefore, property owners do not have to retrofit doors on buildings built before that date. The ADA requires only that owners of public accommodations such as shopping centers, hotels, and restaurants remove access barriers in existing buildings when doing so does not require significant difficulty or expense (significant difficulty or expense being undefined). Businesses may wait to make modifications until they have the financial means to afford the upgrades. The federal government does offer tax credits for retrofits. However, all buildings constructed or modified after the effective date must comply with the law.

▫ If there are stairs at a main entrance, is there also a ramp or lift present?

▫ If a main entrance is not accessible, is there another accessible public entrance to the building?

☐ Do all inaccessible entrances provide directional signage to an accessible entrance?

☐ Is the international symbol of accessibility provided at the accessible entrance?

☐ If an alternate public entrance is used, is it kept unlocked, to provide for independent usage?

☐ Are entryways ≥32 inches wide, as measured between the door's face and the opposite doorstop when the door is open 90°? Notes: (1) Clearance around doors must be ≥36 inches. (2) The law exempts doors that do not require full passage to access a room, such as an entryway into a closet less than 2-feet deep. The minimum width on such doors is 20 inches. (3) For double doors, the minimum width is 48 inches, and the doors must swing in the same direction. At least one door must provide ≥32 inches opening. (4) The floor or ground around the door must be level and clear of obstructions. (5) Doorway thresholds cannot be higher than ¾ inch for exterior sliding doors, and no higher than ½ inch for all other door types. (6) If a door is at the end of a hallway, wheelchair access requires a clearance around the door of ≥54 inches.

☐ Can door handles be operated with one hand and without requiring tight grasping, pinching, or twisting of the wrist? Notes: (1) Hardware must be designed for easy grasping with one hand and cannot require tight pinching or wrist twisting to operate. (2) Hardware for opening must be ≤48 inches above the floor. (3) A door with a closer must take at least three seconds to shut to within 3 inches of the latch. (4) The maximum allowed force necessary to open a door is 5 pounds. The exception is a fire door that must have the minimum force allowed under local zoning codes. (5) Automatic-door units must be low-energy models. They must take at least three seconds to open and cannot require more than 15 pounds of force to stop door movement.

☐ Can wheelchairs cross thresholds? Notes: (1) For a straight on approach, there should be ≥18 inches of clear wall space on the pull side of the door next to the handle for 5'-0" back. (2) A level landing is needed at the accessible door to permit maneuvering and simultaneous door operation.

☐ If the door has a closer, does it take at least 5 seconds to close from a door angle of 90° to 12°?

☐ Are there kick plates 12 inches high extending the width of the door, on the push side, except for automatic and power doors?

☐ Are mats a tripping hazard? Notes: (1) Thin/flexible doormats should be secured to the floor at all edges. (2) A wheelchair should be able to move onto the mats easily.

☐ Where two hinged or pivoted doors are in series, are there ≥48 inches plus the width of the door swinging into the space available between doors.

• Ramps: Any slope greater than 1:20 along an accessible route is considered a ramp. Where space limitations prohibit the use of a 1:12 slope, a ramp may have a slope and rise as follows: (1) A slope between 1:10 and 1:12 is allowed for a maximum rise of 6 inches. (2) A slope between 1:8 and 1:10 is allowed for a maximum rise of 3 inches. (3) A slope steeper than 1:8 is not allowed.

- Do all ramps longer than 6 feet have handrails on both sides? Note: Handrails must be sturdy and 34 inches high.
- Is the width between handrails ≥36-inches?
- Are ramps firm, stable, and non-slip and designed to prevent water buildup on their surfaces?
- Are there level landings located at the top and bottom of all ramps? Note: The landings should have a clear width that is greater than the widest ramp run leading to the landing and the landing's length must be ≥5 feet.
- Where a ramp changes direction, is there a minimum 5-foot by 5-foot landing provided? Notes: (1) Ramps that do not have level landings at changes in direction can create a compound slope that will not meet the requirements of the Department of Justice's *2010 ADA Standards for Accessible Design.* (This is an important document that you should read). (2) Circular or curved ramps continually change direction. Curvilinear ramps with small radii may create compound cross slopes and cannot, by their nature, meet the requirements for accessible routes.
- Do ramps and landings have ≥4-inch edge protection? Note: Edge protection is not needed if a ramp and landing are protected with vertical guardrails or an extended platform that runs ≥12 inches beyond both handrails.
- Are cross slopes[23] of ramp runs no steeper than 1:50?

INTERIOR ACCESSIBLE ROUTES

Per the Department of Justice's *2010 ADA Standards for Accessible Design,* "Accessible routes shall consist of one or more of the following components: walking surfaces with a running slope not steeper than 1:20, doorways, ramps, curb ramps excluding the flared sides, elevators, and platform lifts. All components of an accessible route shall comply with the applicable requirements of Chapter 4" (Accessible Routes).

- Do the accessible entrances provide direct access to the main floors, lobbies, and elevators?
- Are all public spaces on an accessible path of travel?
- Are there 5-foot circles or T-shaped spaces for people who use a wheelchair to reverse direction?
- Are all aisles and pathways to all goods and services ≥36 inches wide? Note: This clear width may be reduced to 32-inches minimum for a length of 24-inches maximum if segments that are 48-inches long minimum and 36-inches wide minimum separate the reduced width segments.
- Is carpeting low-pile, tightly woven, and securely attached along the edges?
- On accessible routes through public areas, are all obstacles cane detectable (located within 27 inches of the floor or protruding less than 4 inches from the wall) or are they higher than 80 inches?
- Do signs designating permanent rooms and spaces, such as restrooms, meeting rooms, and offices comply with the appropriate requirements for accessible signage?

---

[23] Cross slope is the slope (rise over run) of the surface perpendicular to the direction of travel.

- Are all controls that are available for use by the public (including electrical, mechanical, window, cabinet, and self-service controls) located at an accessible height? Note: Reach requirements vary: The maximum height for a side reach is 54-inches and for a forward reach, 48-inches. The minimum reachable height is 15-inches.
- Are doors operable with one hand? Note: Doors should not require tight grasping, pinching, and twisting of the wrist; see Entrance/exterior doors.
- Are Assistive Listening Devices available in meeting rooms, auditoriums, and similar occupancies with a seating capacity of 50 or more people? Notes: (1) The number of receivers/transmitters should be ≥9% of the seating capacity. (2) Assistive Listening Devices are required when reception of audio information by the audience is essential.
- Is there at least one counter for disabled people to use for such activities as check-in, prescription filling, appointment setting, and so forth? Note: Counters should be ≤36-inches?
- With regards to medical facilities:
  ☐ Are there accessible seating spaces in the waiting areas?
  ☐ Is at least one dressing room accessible?
  ☐ Is at least one of each type of testing/treatment room accessible? Note: If the number of rooms of any type exceeds 10, then at least 2 rooms/beds should be accessible.
  ☐ Are all types of medical testing equipment adjustable so as to be accessible and are they located within accessible spaces?
  ☐ Can examining tables and chair heights in accessible treatment rooms be adjusted to or be fixed at ≥17-inches and ≤ 19-inches high?
  ☐ Is there clear floor space parallel and adjacent to examining chairs/tables of at least 30 by 48 inches?
  ☐ Does the interior accessible route extend from the door to the clear floor space?

BATHROOMS

- Are mirrors that are located above lavatories or countertops installed with the bottom edges of their reflecting surface ≤40-inches above the finish floor or ground? Notes: (1) Mirrors that are not located above lavatories should have their reflecting surfaces ≤35-inches above the finish floor or ground. (2) Single full-length mirrors should be usable by people who are ambulatory and people who use wheelchairs. The top edge of mirrors should be ≥74-inches from the floor or ground.
- Are water closets positioned with a wall or partition to the rear and to one side? Notes: (1) The centerline of the water closet should be ≥16 inches to ≤18 inches from the sidewall or partition, except that water closets in the ambulatory accessible toilet compartment should be ≥17 inches and ≤19 inches. (2) Water closets should be arranged for a left-hand or right-hand approach. (3) When the door to the toilet room is placed directly in front of the water closet, the water closet cannot overlap the required maneuvering clearance for the door inside the room. (4) The seat height of a water closet above the finish floor should be ≥17-inches and ≤19-inches measured to the top of the seat.

(5) Seats shall not be sprung to return to a lifted position. (6) Grab bars should be provided on the sidewall closest to the water closet and on the rear wall. (7) Flush controls may be hand operated or automatic. Hand-operated flush controls in wheelchair-accessible toilet compartments should be on the open side of the water closet. (8) Toilet paper dispensers should be ≥7-inches and ≤9 inches in front of the water closet measured to the centerline of the dispenser. The outlet of the dispenser shall be ≥15-inches and ≤48 inches maximum above the finish floor. Dispensers should not be located behind grab bars. Dispensers should not control delivery or continuous paper flow.[24]

- Are lavatories and sinks installed with so that the front of the higher of the rim or counter surface is ≤34-inches above the finish floor or ground?
- Do hand-operated metering faucets remain open for ≥10 seconds?
- Are water supply and drain pipes that are under lavatories and sinks insulated or configured to protect against contact. Note: There should be no sharp or abrasive surfaces under lavatories and sinks.

PAY OR PUBLIC TELEPHONES

With the high use of cell phones, pay telephones have become almost obsolete. Nevertheless, where they do exist, they should comply with ADA.

- If pay or public phones are provided, is there clear floor space of at least 30 by 48 inches in front of at least one?
- Is the highest operable part of the phone no higher than 48-inches (up to 54-inches if a side approach is possible)?
- Do the phones protrude no more than 4-inches into the circulation space?
- Does the phone have push-button controls?
- Is the phone hearing-aid compatible?
- Is the phone adapted with volume control?
- Is one of the phones equipped with a text telephone (TT or TDD)?
- Is the location of the TT identified by accessible signage bearing the international TDD symbol?

ELEVATORS AND LIFTS

See also High-rise Building Construction, Final Inspection—High-rise Building Construction, Elevators, ADA.

- Are elevators equipped with both visible and verbal or audible door opening/ closing and floor indicators (one tone = up, two tones = down)?
- Are the call buttons in the hallway ≤42-inches?
- Do the controls inside the cab have raised and Braille lettering?
- Is there a sign on the jamb at each floor that identifies the floor in raised and Braille letters?

---

[24] The above checklist and notes that pertain to water closets are not complete. If this issue is important for your, due diligence/capital projects inspection, e.g., if you are inspecting a medical/treatment facility or a facility that is used by children, refer to "ADA Design Guide," U.S. Department of Justice, Civil Rights Division, *Disability Rights Section.*

- Are all emergency intercoms usable without voice communication?
- Are controls between ≥15-inches and ≤ 48-inches high (≤54-inches if a side approach is possible)?
- Are elevator cabs ≥48 by ≥48 inches? If not, can usability be demonstrated?
- Do all elevator doors provide a clear opening ≥36-inches wide?
- Are car controls ≥36-inches to ≤ 54-inches high?
- Lifts: Can the lift be used without assistance?
- Lifts: Is there ≥30-inches by ≥48-inches of clear space for a person using a wheelchair to approach, reach the controls, and use the lift?
- Lifts: Are controls between ≥15-inches and ≤ 48-inches high (≤54-inches if a side approach is possible)?

EMERGENCY EGRESS

- Do all alarms have both flashing lights and audible signals?
- At minimum, are visual signal appliances provided in buildings and facilities in each of the following areas: restrooms and any other general usage areas (e.g., meeting rooms), hallways, lobbies, and any other area for common use?

STAIRS

- Do treads have a non-slip surface?
- Do stairs have continuous rails on both sides, with 12-inch extensions beyond the top and bottom stairs?

DRINKING FOUNTAINS

Where drinking fountains are available for public use, the following criteria should be met: the spouts should be ≤36-inches; spouts should be located toward the front of the unit; the water flow should run parallel to the front of the unit or ≤3-inches of the front edge; both front-mounted and side-mounted controls should be near the front edge; there should be a clear floor space of ≥48-inches wide and ≥30-inches deep (measured from the front edge of the fountain); and there should be a clear knee space of ≥27-inches, measured from the bottom of the apron to the floor (wall mount-cantilevered units). If the fountain is located in an alcove, there should be ≥30-inches clear width (wall mounted-cantilevered units). Wall-mount-cantilevered units should be ≥17-inches to ≤19 inches deep and there should be a clear toe space of ≥9 inches, measured from the bottom of the fountain to the floor.

HEATING, VENTILATING, AND AIR CONDITIONING SYSTEMS (HVAC)

Building heating, ventilating, and air conditioning systems have a major impact on a user's perception of the building. If not operating properly, or designed properly in the first place, it will also adversely affect users' performance and even their health. The system must maintain the inside air at the proper temperature, humidity, and quality. The system should be sensitive to esthetic concerns such as velocity of drafts, noise levels, and appearance. Ideally, a significant number of daily users should be able to control their individual spaces.

A HVAC system has two important influences on the building. First, users expect to work in conditions that are conducive to their health, safety, and productivity. Second, the energy and maintenance costs of operating the HVAC system are substantial; they consume a significant percentage of buildings operating costs. HVAC systems are also costly to replace; they typically represent 9% to 16% of the cost of constructing a new building.

Typically due diligence/capital project planning inspections include the following components: boilers, radiation, and solar heating, if present; ductwork and piping; fans; heat pumps; fan coil units; air handling units; rooftop A/C; water chillers; cooling towers; and computer rooms and other specialty cooling.

INVENTORY AND DESCRIPTION

An inventory should be made of all major pieces of equipment, to include their description, manufacturer, age (date of installation), size, and location. Prior to your inspection, you may set criteria, for example, age or type that will automatically be identified for replacement. Include the following items, if present, and rate them as best possible:

- Air conditioning:
  - Chilled water piping.
  - Chillers, centrifugal water cooler.
  - Chillers, absorption.
  - Chillers, reciprocating, air cooled.
  - Chillers, scroll/rotary, air cooled.
  - Chillers, reciprocating, water-cooled.
  - Chillers, emergency stop switch.
  - Chillers, scroll/rotary, water cooled.
  - Cooling towers.
  - Cooling tower pumps.
  - Air-cooled condensers.
  - Dx_x air-cooled condensing units.
  - Fans, return.
  - Package/rooftop units.
  - Split heat pumps.
  - Other.
- Climate control:
  - Temperature control air compressors.
  - Temperature control sub-panels.
  - Temperature control pneumatic sub-panels.
  - Temperature control pneumatic valves.
  - Temperature control-refrigerated air driers.
  - Temperature control steam traps.
  - Other.
- Direct digital control system (DDC):
  - Pneumatic tubing.
  - Temperature control thermostats.
  - Temperature control pneumatic pressure=reducing valves.

       ☐   Other. Notes: (1) Identify the manufacturer and the sophistication of the system. (2) Check to see if the system is actually being used or if the facilities staff has placed the system on manual and is walking around with thermometers in their hands. (3) Speak with people in the control room and get their opinion of the system and the support they are receiving from the manufacturer/installer.

- Heating:
  - ☐ Expansion tanks.
  - ☐ Heating surfaces, commercial-type convectors.
  - ☐ Heating surfaces, fin tube.
  - ☐ Heating surfaces, heating coils.
  - ☐ Hot water heat exchangers, including motors, and motor starters.
  - ☐ Pressure-reducing valves.
  - ☐ Steam and condensate piping.
  - ☐ Other.
- Heating plants:
  - ☐ Boiler auxiliaries:
    - Boiler blow down/piping systems.
    - Boiler breeching.
    - Boiler dampers.
    - Boiler emergency stop switches.
    - Boiler feedwater systems.
    - Boiler feedwater treatment.
    - Boiler leader piping/valves.
    - Boiler safety valves.
    - Piping, water line (makeup).
    - Steam condensate return gravity systems.
    - Steam condensate return vacuum systems.
    - Motors/motor starters.
    - Other.
  - ☐ Boiler Systems:
    - Fuel used: coal, gas, oil, hot water, steam.
    - Boiler control system.
    - Burner control panel.
    - Combustion air system.
    - Gas meter-room exhaust system.
    - Fuel oil storage.
    - Oil leak detectors.
    - Fuel oil–level gauges.
    - Overfill alarms.
    - Fuel oil piping.
    - Fuel oil pump heater sets.
    - Fuel tank heaters.
    - Fuel transfer pumps.
    - Fuel transfer pump motors/starters.
    - Gas leak detection systems.
    - Other.

SYSTEM INSPECTIONS AND EVALUATIONS

Many of the following questions in this section are USGBC (LEED) driven.

- General:
  - ☐ Check and comment on the overall cleaning, maintenance, and repair of registers, grills, dampers, louvers, bird and insect screens, supply and return ducts drain pans, coils, bag collection, wet collectors, fire hazards, and so forth.
  - ☐ Have planned periodic maintenance programs been in place? Notes: (1) Review receipts for belts, filters, grease, and other consumable items. (2) Check to see if a logbook has been kept to verify routine maintenance. (3) Check lubrication of bearings and moving parts; listen for unusual noises. (4) Belts should not be worn or frayed.
  - ☐ Have valves been exercised on a periodic basis? Have safety valves been lifted periodically?
  - ☐ Have high-pressure relief valves been lab tested annually and low-pressure relief valves tested every five years?
  - ☐ Are all systems free of rust and corrosion?
  - ☐ Are wiring and electrical controls in satisfactory condition?
  - ☐ Are all automatic control systems working properly and not being bypassed by the operations staff.
  - ☐ Are all thermal insulation and protective coatings free of excessive wear?
  - ☐ Are all guards, casings, hangers, supports, platforms, railings, and mounting bolts secure?
  - ☐ Are piping and piping systems properly identified?
  - ☐ Are solenoid valves functioning properly?
  - ☐ Are burner system assemblies, combustion chambers, and smoke pipes functioning properly?
  - ☐ Are electrical heating units functioning properly?
  - ☐ Are all steam and hot water heating equipment functioning properly?
  - ☐ Are accessible steam, water and fuel piping is satisfactory condition?
  - ☐ Are traps functioning properly? Note: An easy check is to go to the roof and look for plumes.
  - ☐ Are humidifier assemblies functioning properly?
  - ☐ Are water sprays, weirs, and similar devices functioning properly?
  - ☐ Are shell-and-tube type condensers functioning properly?
  - ☐ Are self-contained evaporative condensers functioning properly?
  - ☐ Are air-cooled condensers functioning properly?
  - ☐ Are all compressors functioning properly?
  - ☐ Are liquid receivers functioning properly?
  - ☐ Are refrigerant driers, strainers, valves, oil traps, and ancillary equipment functioning properly?
  - ☐ Is a water treatment program in place for boilers and cooling towers? Notes: (1) Weekly water testing should include pH, hardness, and alkalinity. Also total dissolved solids should be included for cooling towers and high-pressure boilers. (2) Check the credentials of the people who test

the water and add chemicals to it; have they been properly trained or are they "winging it?"

- ☐ Are air handlers free of mold?
- Outside air introduction and exhaust systems:
  - ☐ Do the HVAC systems provide sufficient capacity to sustain long-term occupant comfort and well-being? Note: Usually, HVAC systems should not have too much capacity thereby causing them to cycle on and off on all but the hottest days.
  - ☐ Are primary HVAC systems and assemblies performing as intended?
  - ☐ Do the system supply-side outdoor air ventilation rates meet the criteria set by ASHRAE 62.1-2004? Notes: (1) If this is not feasible due to physical constraints, determine if the system in question can be modified to provide at least 10 CFM per person. (2) This should be commented on in the inspection report.
  - ☐ Per LEED requirements, have permanent monitoring systems that provide feedback on the ventilation systems' performance been installed?
  - ☐ Do mechanical ventilation systems that predominantly serve densely occupied spaces[25] have CO_2[26] sensors or sampling devices installed?
  - ☐ Have CO_2 sensors been tested and calibrated to an accuracy of no less than 75 ppm or 5% of the reading, whichever is greater?
  - ☐ Have CO_2 sensors been monitored by a system that is capable of and configured to trend CO_2 concentrations on no more than 30-minute intervals?
  - ☐ Are systems configured to generate an alarm visible to a system operator or building occupants if the CO_2 concentrations in any zone rises more than 15% above that corresponding to the minimum outdoor air rate required by ASHRAE Standard ASHRAE 62.1-2004?
  - ☐ Are the CO_2 sensors for demand-controlled ventilation that comply with ASHRAE 62.1-2004?
- Other mechanical ventilation systems:
  - ☐ Have outdoor airflow measurement devices been provided? Note: The measuring devices should be capable of measuring (and, if necessary controlling) the minimum outdoor airflow rate at all expected system operating conditions within 15% of the design minimum outdoor air rate.
  - ☐ Is each outdoor airflow measurement device monitored by a control system that is capable of and configured to trend outdoor airflow on no more than 15-minute intervals?
  - ☐ Is each control system capable and configured to generate an alarm that is visible to the system operator if the minimum outdoor air rate falls more than 15% below the design rate?

---

[25] Densely occupied space: An area that has a design occupant density of 25 people or more per 1,000 square feet (40 square feet or less per person). Most codes specify ventilation requirements for these spaces.

[26] A carbon dioxide (CO_2) sensor: An instrument for the measurement of carbon dioxide gas. Carbon dioxide is an important measure in monitoring <u>indoor air quality</u> and many industrial processes.

- Natural ventilation systems:
  - ▢ Are there CO_2 sensors located in the breathing zone of every densely populated room?
  - ▢ Are there CO_2 sensors located in the breathing zone of every natural ventilation zone?
  - ▢ Are there CO_2 sensors located outdoors?
  - ▢ Do CO_2 sensors provide an audible or visual alarm to the occupants in the space and building management if CO_2 conditions are greater than 530 ppm above outdoor CO_2 levels or 1,000 ppm absolute?
  - ▢ Do operable window areas meet the requirements of ASHRAE 62.1-2004, section 5.1?
  - ▢ Have efforts been made to protect building occupants and maintenance personnel from potentially hazardous particle contaminants?
  - ▢ Are filters with particle removal effectiveness MERV13 or greater in place at all outside air intakes and at the returns for the re-circulation of inside air[27]?
  - ▢ Are filters replaced on a schedule?
- Environmental tobacco smoke (ETS) control: Some owners may not be interested in this topic and, therefore, it may be omitted from an inspection. However, smoking has been found to adversely affect those who are near smokers, and more and more smoking indoors is becoming illegal.
  - ▢ Has building management taken steps to minimize the adverse effects of environmental tobacco smoke (ETS) on building occupants, indoor surfaces, and building systems? (As an aside, ETS adversely affects building occupants, indoor surfaces and building systems.)
  - ▢ Is smoking prohibited within the building?
  - ▢ Are exterior designated smoking areas at least 25 feet away from building entrances, outdoor air intakes, and operable windows?
  - ▢ Are designated smoking rooms designed to contain, capture, and remove ETS from the building? Note: Smoking areas should exhaust directly outdoors, away from air intakes and building entries, and deck-to-deck partitions. They should beat negative pressure.
- Controllability of systems—temperature and ventilation[28]:
  - ▢ Does the systems provide for high levels of temperature and ventilation control by individual occupants or specific groups in multi-occupied spaces, for example, conference rooms? Notes: (1) As a marker, check if at least 50% of the occupants can adjust temperature and ventilation to suit their individual needs. (2) Operable windows may be considered in lieu of individual controls for occupants in spaces that are ≤20-feet inside of and ≤10-feet to either side of the operable part of the window.

---

[27] Reference: ASHRAE Standard 52.2-1999: Method of Testing General Ventilation Air-Cleaning Devices for Removal Efficiency by Particle Size (ANSI Approved).

[28] References: (1) ASHRAE 62.1-2004, paragraph 5.1. (2) ASHRAE Standard 55-2004, Thermal Comfort Conditions for Human Occupancy.

- Ozone Protection[29]: Check to learn if any HVAC, refrigeration, or fire suppression systems contain CFCs, HCFCs, or Halons?
- System maintenance[30]:
  - ☐ Has a documented preventive maintenance program that, at a minimum, includes regular sensor and actuator calibration for the following components been in place?
  - ☐ Outside air temperature;
  - ☐ Mixed air temperature;
  - ☐ Return air temperature;
  - ☐ Discharge or supply air temperature;
  - ☐ Coil face discharge air temperature;
  - ☐ Chilled water supply temperature;
  - ☐ Condenser entering water temperature;
  - ☐ Heating water supply temperature;
  - ☐ Wet bulb temperature or relative humidity sensors;
  - ☐ Space temperature sensors;
  - ☐ Economizer and related dampers;
  - ☐ Cooling and heating coil valves;
  - ☐ Static pressure transmitters;
  - ☐ Air and water flow rates; and
  - ☐ Terminal unit dampers and flows?

PLUMBING

As Americans, we continue to increase water usage. The U.S. Geological Survey estimates that between 1990 and 2000, public water supply use increased 12% to 43.3 billion gallons per day. This high demand is straining supplies, so it is important that our inspections identify how water is being used and whether it is being used efficiently and wisely; note any obvious water wasteful practices. Also if the facility is regulated by the EPA National Pollution Discharge Elimination System (NPDES) Clean Water Act requirements inspectors need to ensure that the facility in compliance with the act?

Generally, primary components of interest for due diligence/capital project planning inspections are sanitary, storm water and irrigation systems; plumbing fixtures; piping and meters.

OCCUPIED SPACES

- Is water supply adequate and is pressure adequate?
- Are there any obvious sanitation hazards?
- Are drains flowing properly?
- Are fixtures attractive and functioning properly?
- Are fixtures available for disabled people? Note: See ADA checklists.
- Is there corrosion on pipes and fittings?
- Are hangers and clamps secure?

---

[29] Reference: U.S. EPA Clean Air Act, Title VI, Section 608.
[30] Reference: USGBC, Existing Buildings: Operations & Maintenance, Reference Guide.

- Does water have a foul odor or taste? Note: Check to see that the water has been tested per EPA requirements.
- Are main cutoffs operable?
- Are there sufficient cutoff valves to isolate areas logically for maintenance and emergencies (without shutting down the entire facility)?
- Are hot water settings proper (not too hot or cold)?
- Are pumps operating properly?
- Is pipe insulation in good condition?
- Have flow reduction aerators been installed where required?
- Have automatic controls on lavatories been installed?
- Have high-efficiency urinals and toilets been installed?
- Have high-efficiency showerheads and kitchen faucets been installed?
- Has a greywater system for flushing been installed?

SANITARY SYSTEMS

Most facilities send their sewerage to a public sewage treatment plant. If this is not the case special mention should be made of any on-site wastewater treatment including conventional biological treatment facilities or natural processes that treat wastewater. Inspectors may need additional professional advice to inspect such facilities.

- Is flow adequate?
- Are there sufficient cleanouts? Notes: (1) Normally, cleanouts should not be more than 50 feet apart in horizontal drainage lines. (2) Cleanouts should be in compliance with the local code and in your opinion large enough.
- Do bathrooms, janitors' closets, boiler rooms, and so forth, have floor drains and are they connected to the sanitary system?
- Where wastewater flows by gravity, does it flow as designed?
- Where wastewater is pumped, have the pumps been well maintained?
- Are there accessible backwater valves wherever fixtures, floor drains, or area drains are subject to overflow from the public sewer system? Note: Most codes direct where backwater valves (backflow preventers) are required.

STORM WATER SYSTEMS

Storm water is precipitation from rain and snowmelt that flows over land and impervious surfaces; storm water does not percolate into the ground. As it flows, it accumulates debris, chemicals, sediment, or other pollutants that could adversely affect water quality if best management practices (BMPs) are not followed.

Most storm water discharges are considered point sources and require coverage under an EPA-administered *NPDES* permit. Most states are authorized to implement the Stormwater NPDES permitting program. EPA remains the permitting authority in a few states, territories, and on most land in Indian Country. If storm water is a major issue, for example, if there are stringent local requirements in place to protect natural habitat, waterways, and water supplies from pollution, you may have to make your inspection more detailed than covered here; you can obtain additional information by contacting EPA. However, for most due diligence/capital planning program inspections, a review of the section on Grading, Drainage, and Landscaping should provide enough information for your inspection. Usually spelled as stormwater, not storm water.

Inspectors should comment as appropriate, especially about harvested storm water in tanks/cisterns and the reuse of water for irrigation and flushing of toilets and urinals.

IRRIGATION SYSTEMS

Compare the irrigation system of the facility to others in the area. If a water savings system has been installed, describe the system and estimate the water savings compared with the commonly used system. Look for obvious leaks.

PLUMBING FIXTURES

Usually due diligence/capital projects planning inspections include a narrative that describes the plumbing fixtures and fittings within the facilities (restrooms, kitchens, janitorial closets, etc.). Do they comply with the latest Uniform Plumbing Code (UPC) and International Plumbing Code (IPC) fixture and fitting performance requirements or an overriding local code.

FIXTURE	UPC AND IPC STANDARDS FOR PLUMBING FIXTURE WATER USE	WATER SENSE[†] STANDARDS
Water closets (gpf)	1.60	1.28
Urinals (gpf)	1.00	0.50a
Shower heads(gpm)*	2.50	1.50-2.00b
Public lavatory faucets and aerators (gpm)**	0.50	-----
Private lavatory faucets and aerators (gpm)**	2.20	1.50
Public metering lavatory faucets (gpm/metering cycle)	0.25	-----
Kitchen and janitor sink faucets	2.20	-----
Metering faucets (gal/cycle)	0.25	-----

*When measured at a flowing water pressure of 80 psi.

**When measured at a flowing water pressure of 60 psi.

[†]WaterSense is a partnership program sponsored by EPA. Its goal is to protect the future of the U.S. water supply by promoting and enhancing the market for water-efficient products and services. WaterSense is a voluntary program. EPA develops specifications for water-efficient products through a public process. If a manufacturer makes a product that meets those specifications, the product is eligible for third-party testing. If the product passes the test, the manufacturer can put the WaterSense label on that product.

a. On May 22, 2008, EPA issued a notification of intent to develop a specification for high-efficiency urinals. WaterSense anticipates establishing a maximum allowable flush volume of 0.5 gpf.

b. On August 30, 2007, EPA issued notification of intent to develop a specification for showerheads. WaterSense anticipates establishing a single maximum flow rate between 1.5 and 2.0 gpm.

In conducting due diligence/capital planning program inspections, you should have eye on whether the current facilities staff has strived to reduce indoor fixture and fitting water use. If the owner is interested in LEED certification, your inspection will of necessity become more detailed and will require significant amount of calculations; you may wish to bring in a LEED-accredited professional to assist with these calculations.

You may also suggest to the owner the development and implementation of a policy requiring an economic assessment of conversion to high-performance plumbing fixtures and fittings as part of any future indoor plumbing renovation. Such an assessment would account for potential water supply and disposal cost savings and maintenance cost savings. If the owner wants to pursue LEED certification, ask the following questions:

- Has potable water usage[31] of indoor plumbing fixtures and fittings been reduced to a level equal to or below the LEED for Existing Buildings: O&M baseline? Notes: (1) Calculations should assume that 100% of the building's indoor plumbing fixtures and fittings meet the UPC 2012 or IPC 2012 fixture and fitting performance requirements. Fixtures and fittings included in the calculations are water closets, urinals, showerheads, faucets, faucet replacement aerators, and metering faucets. (2) The LEED for Existing Buildings: O&M baseline water usage is set depending on the year of substantial completion of the building's indoor plumbing system. Substantial completion is defined as either initial building construction or the last plumbing renovation of all or part of the building that included 100% retrofit of all plumbing fixtures and fittings as part of the renovation.
- Set the baseline as follows:
  - For a plumbing system substantially completed in 1994 or later throughout the building, the baseline is 120% of the water usage that would result if all fixtures met the codes cited above.
  - For a plumbing system substantially completed before 1994 throughout the building, the baseline is 160% of the water usage that would result if all fixtures met the codes cited above.
  - If indoor plumbing systems were substantially completed at different times for different parts of the building because the plumbing renovations occurred at different times, set a whole-building average baseline by prorating between the above limits. Prorate based on the proportion of plumbing fixtures installed during the plumbing renovations in each date period, as explained in the latest LEED for Existing Buildings: Operations &Maintenance Reference Guide. Pre-1994 buildings that have had only minor fixture retrofits (aerators, showerheads, flushing valves) but no plumbing renovations after 1993 may use the 160% baseline for the whole building.
  - Demonstrate fixture and fitting performance through calculations to compare the water use of the as-installed fixtures and fittings with the use of UPC- or IPC-compliant fixtures and fittings, as explained in the latest LEED for Existing Buildings: Operations &Maintenance Reference Guide.
- Has an effort been made to identify water usage through metering with an eye toward long-range water conservation?

---

[31] Potable water is defined as water that is suitable for drinking and is supplied from wells or municipal water systems.

☐   Do facilities have permanently installed water meter(s) that measure total potable water use for the entire building(s) and associated grounds?

☐   Do water meters serve at least 80% of the irrigated landscaped area?

☐   Do water meters serve at least 80% of the indoor plumbing fixtures and fittings?

☐   Is replacement water for water towers metered?

☐   Do water meters serve at least 80% of the installed domestic hot water heating capacity (including both tanks and on-demand heaters)?

☐   Do water meters serve at least 80% of expected water consumption for process-type end uses, for example, humidification systems, dishwashers, clothes washers, pools, and other systems using process water?

GAS PIPING

- Do the buildings have an outside gas service line valve or other outside emergency shut-off device or method that is acceptable to the local fire marshal? Note: Automatic gas shut-off is a requirement in earthquake zones.
- Is outside gas service piping adequately protected from the elements and vehicular traffic?

GAS METERS

- Are gas meters located as near as practicable to the point of entrance of the service? Note: Inspectors should identify meter locations in their reports.
- Are there separate building natural gas meters that allow aggregation of all process natural gas loads?
- Have separate meters that allow aggregation of all indoor occupants' fixture water usage been installed?
- Have separate meters that allow aggregation of all indoor process water usage been installed?

ADDITIONAL CONSIDERATIONS

Inspectors should comment as appropriate.

ELECTRICAL

We are totally dependent on electrical power. So much so that when we lose it we are, for all intents and purposes, out of business. Major outages are rare but have occurred. And when they did, whole regions were out of business. At the microlevel of a facility, everything must be done to ensure reliable, clean power and in the event of an emergency, sufficient emergency power to keep vital functions operating.

The primary components of interest are service and distribution, wiring, communications and security systems, emergency power, exterior and interior lighting, and if applicable on-site and off-site renewal energy.

SERVICE AND DISTRIBUTION

Inspection reports should include general description of the electrical service and distribution system including, but not limited to, amperage, general condition, types of transformers, average watts per square foot, and whether there are any hazardous conditions, which should be reported immediately. Note also whether the feed source is owner or utility owned and above or below ground.

- General:
  - □ Are line drawings available?
  - □ Are systems grounded in accordance with the National Electric Code?
  - □ Do all buildings have lightning protection?
- Motor control centers: Motor control centers (MCCs) are modular cabinet systems for powering and controlling motors, usually in a factory having heavy machinery. Infrared inspection is an effective tool for inspecting them.
  - □ Is there excessive heat on the surface of doors? Note: If the condition is unacceptable, do not open the door.
  - □ Are there abnormal thermal patterns caused by high-resistance connections, overloads, or load imbalances?
  - □ Are the connections to the main in satisfactory condition? Note: These may be difficult to see.
  - □ Do motor starters match the voltage and horsepower of the system? Note: Motors may be damaged or the life reduced if they have been operated at a current above full-load current; there should be motor-overload protection.
- Transformers: As transformers age, gaseous and liquid contaminants become dissolved in the dielectric fluid. These may lead to shortened useful lives and even failure. Sampling and analysis of dielectric fluid is the cornerstone of your transformer inspection. Transformer oil should be sampled and analyzed using gas chromatographs, spectrometers, and other sophisticated test equipment. The results of these tests show the types and amounts of contaminants in the oil and how they have changed from new. Unless you consider yourself expert in transformers, you will want someone who is an expert to analyze the data from the test sample.

    Identify who owns the transformers: utility or facility.

    Assuming facility ownership, record nameplate and other operational data and do a complete visual inspection of the transformer at the time the samples are drawn. Check if the transformers have been tested on a regular basis. Check also if the transformers contain PCBs.
  - □ Is there evidence of leaks, inoperative gauges, cracked bushings, low oil level, and other defects? Note: Look for signs of arching and burning.
- Wiring, wall switches, lighting fixtures: Identify the types of lighting fixtures, fluorescent, HID, or incandescent, and record their general condition. Estimate the numbers of fixtures.
  - □ Is there a program for properly disposing of lights in an environmentally sound manner? Note: Check to see if the current building owner has taken measures to minimize the amount of mercury brought into the building site through purchases of lamps. LEED assigns one point when mercury is at or below 90 picograms[32] per lumen-hour and two points when mercury is at 70 picograms per lumen-hour.
  - □ Are cover plates in place?
  - □ Are junction boxes covered?

---

[32] Picogram: One trillionth (10^{-12}) of a gram.

- ☐ Is all wiring in conduits?
- ☐ Are sufficient outlets provided?
- ☐ Are GFI circuit breaks where required?
- ☐ Are all switches operational?
- ☐ Is the electrical system balanced? Note: An electrical engineer may have to help you answer this question.
- ☐ Do lights make buzzing/humming noises?
- Communications and security:
  - ☐ Is a public address system installed? Is it operable? Can it be heard throughout the building?
  - ☐ Is there a central control panel/intrusion alarm?
  - ☐ Is there a closed-circuit television system?
  - ☐ Is there an infrared or ultrasonic intrusion alarm?
  - ☐ What type of telephone system is installed?
  - ☐ Is TV cable or satellite installed?
- Emergency power:
  - ☐ Has an emergency generator been installed? Is it diesel or natural gas? Note: Somehow there is never one right location for an emergency generator: basements flood and exhausts waffle up through the building, top floor generators are inconvenient to service and refuel and their noise may be a problem. Comment based on your experience for the specific facility.
  - ☐ Provide data plate information, e. g., manufacturer, age, and so forth.
  - ☐ Is there a logbook of maintenance/testing?
  - ☐ Is system periodically tested under load? Are switches exercised regularly?
  - ☐ Does the system have an automatic transfer switch?
  - ☐ Are there problems with cooling, exhaust, or noise?
  - ☐ What is fuel storage capacity and is tank topped off?
  - ☐ Have emergency/exit lights been maintained?
  - ☐ Is emergency/exit lighting tied into emergency generator or batteries?
  - ☐ Is emergency DC standby battery power installed (lead-acid or nickel-cadmium bank)? Note: While DC power is mostly a thing of the past, there may be some old equipment such as lifts that use DC.

EXTERIOR AND SITE LIGHTING

Light pollution from poorly designed outdoor lighting affects the nocturnal ecosystem on the site and adversely affects nearby neighbors without improving security/safety. Excessive outdoor lighting wastes electricity and money. The questions in this section are based primarily on the requirements for a LEED certification.

- Light to night sky:
  - ☐ Is exterior site lighting used primarily for security and safety? Note: Lighting that is used solely to highlight architectural features should be identified; it is frequently unnecessary.
  - ☐ Are downlighting[33] rather than uplighting[34] techniques used?

---

[33] Downlighting: Light that is cast downward from a fixture. Down lighting is the most common and direct form of lighting. Downlighting is also spelled "down lighting."

[34] Uplighting: A source of light that is cast upward that is usually used to illuminate a ceiling cavity or architectural feature for esthetic reasons. When combined with reflective ceiling materials, uplighting can function as a source of indirect lighting. Uplighting is also spelled "up lighting."

□ Are full cutoff luminaries used?

□ Are all outdoor luminaries 50 W and over shielded so that they do not emit light to the night sky??

- Light trespass:

□ Have the night illumination levels been measured at regularly spaced points around the perimeter of the property? Notes: (1) Measurements should be taken with the building's exterior and site lights both on and off. (2) For LEED certification, at least eight measurements are required at a maximum spacing of 100 feet. The illumination level measured with the lights on must not be more than 20% above the level measured with the lights off. This requirement must be met at each measuring point.

INTERIOR LIGHTING

The questions in this section are based primarily on the requirements for a LEED certification. Comment as appropriate.

- Controlled lighting:

□ What percent of individual workstations can control lighting?

□ What percent of multioccupied spaces can be controlled by users?

□ Are all (or some) nonemergency built-in lighting with a direct line of sight to any openings in the building envelope (translucent or transparent, wall or ceiling) automatically controlled to turn off during all after-hour periods?

□ How many off-hours have been programmed annually? Is there an override for after-hours work?

- Daylighting[35]: This is an important issue for owners who are interested in obtaining a LEED certification; a building may earn up to two LEED points if it achieves a 2% daylight factor (DF) in 75% of all spaces occupied for critical visual tasks or if it achieves direct line of sight vision glazing for building occupants in 90% of regularly occupied spaces (one point for 50 or 45%). The DF is the ratio of exterior illumination to interior illumination expressed as a percentage.

LEED considers rooms with a total daylighting factor of 2.0 percent or greater as being daylit. Once daylighting factors have been calculated for all regularly occupied spaces (areas where workers are seated or standing as they work inside a building), the sum of regularly occupied space and the sum of day lit space is determined. The total daylit percentage is then calculated by dividing the total daylit area by the totally regularly occupied area.

For LEED credits, an exacting calculation is required. However, it is possible to estimate

A simple rule of thumb can be used to approximate the DF:

$$DF = 0.1 \times P$$

where, DF = daylight factor
P = Percentage glazing to floor area, for example, given a room of 100-m^2 floor area with 20 m^2 of glazing

[35] Daylighting: The illumination of buildings by natural light. Daylighting is also spelled "day lighting."

$$DF = 0.1 \times (20 \div 100) \times (100 \div 1) = 2\%$$

This can be more usefully represented in calculation of the natural illuminance[36] at the reference point inside a building by applying the following formula:

$$DF = (Ei \div Eo) \times 100$$

where, DF= Daylight factor
Ei = Illuminance at reference point in building
Eo = Illuminance at the reference point if the room were unobstructed

Both factors of E are measured in lumens per square meter (lux), with Eo taken as a standard 5,000 lux for unobstructed sky in the United Kingdom. So transposing the formula to make Ei the subject,
$Ei = (DF \times Eo) \div 100$
$Ei = (2 \times 5000) \div 100 = 100$ lux

Daylight readings at a reference point in a room can be made up of three components:

- ☐ Sky component or the light received directly from the sky
- ☐ Externally reflected component, which is the light received after reflection from the ground, building, or other external surface, and
- ☐ Internally reflected component, which is the light received after being reflected from the surfaces inside a building
- • On-site and off-site renewal energy: LEED encourages and recognizes increasing levels of on-site and off-site renewal energy to reduce environmental impacts associated with fossil fuel energy use. If such a program is in place, elicit from the owner a summary of the program, for example photovoltaic, biofuel-based electrical system, purchased renewable energy certificates. Comment as appropriate.

FIRE AND SAFETY

Fire protection is a major concern in all facilities. While the checklists that are presented focus primarily on installed items, of equal importance is employee training that deals with proper actions to take to reduce the likelihood of fires and how to respond if one should occur; depending on the owners wishes this aspect of fire and safety may not be included in your due diligence inspection.

---

[36] Illuminance: The luminous flux incident on unit area of a surface. It is measured in lux (the International System unit of illumination, equal to one lumen per square meter). Lumen: The unit of luminous flux in the International System, equal to the amount of light given out through a solid angle by a source of 1-candela intensity radiating equally in all directions. Candela: A unit of luminous intensity equal to $1/_{60}$ of the luminous intensity per square centimeter of a blackbody radiating at the temperature of solidification of platinum (2,046°K). Also called candle.

Fire and safety systems include detection notification, suppression, and smoke control systems of various types. Detection devices include smoke, CO, and heat detectors. Notification systems make use of audible alarms, visual alarms, and telephone connection to the fire department. Suppression equipment includes portable extinguishers, kitchen hood systems, and sprinklers. Smoke control relies on fans, vents, and dampers. Local codes, regulations, and standards usually dictate which of these components are installed.

ELECTRICAL

- Electrical—general: Obviously, if circuit breakers are regularly tripping, there is an unsatisfactory condition that should be reported.
  - Do electrical panels have voids or open spaces?
  - Are all electrical panels easily accessible?
  - Are all electrical panels labeled with a notation for each breaker? Note: This is often overlooked. It is also worth checking to see if the labels are correct.
  - Are all cords and plugs in good condition? Notes: (1) Extension cords should not be run under carpets/rugs or across doorways. (2)Check to see if the facility has a policy in place that requires the facility maintenance department to approve all extension cords
  - Are all ground plugs in good condition?
  - Are all electrical boxes equipped with an undamaged cover plate?
  - Are electrical switches, switch plates, or receptacles is good condition? Note: Look especially for cracked, broken, or exposed contacts.
  - Are all electric panels locked and do surrounding spaces have ≥3-feet of clear space?
  - Are electrical circuit panels breaker clearly identified?
  - Do all electrical outlets that are ≤6-feet of a sink or basin have GFCI-type outlets?
- Emergency lighting:
  - Do all egress pathways have emergency lighting?
  - Have all emergency lighting units been tested for 30 seconds in the past 30 days?
  - Are all corridor nightlights lit at night?
  - If power failure occurred within the last 30-days, did the emergency lighting come on?

HAZARDOUS MATERIALS

- Are Material Safety Data Sheets available on all hazardous materials used within the facility?
- Are Material Safety Data Sheets readily available for staff in areas where chemicals are used/stored?
- Are Emergency Measures posted in case of accidental spills?
- Are Oxygen warning signs posted where Oxygen is stored?
- Does the facility contain ≤10 gallons (total collective) of Class I, II, or III flammable liquids?

- Are flammable liquids stored in proper containers? Note: Generally, both the amounts and storage cabinets that are used for chemical/flammable liquids storage are regulated. Inspectors should check with local fire department officials and codes.
- Is personal protective equipment available for staff who use flammable liquids/chemicals? Has staff been trained?
- Are safety storage cabinet vents clear of obstructions?
- Are soiled rags kept in approved self-closing waste containers?

FIRE SAFETY

- Fire alarm systems:
  - ☐ Does the fire alarm control panel indicate that it is operational?
  - ☐ Are a portion of the smoke detectors tested in each alarm circuit semi-annually?
  - ☐ Have all smoke detectors been tested within the past 12 months?
  - ☐ Are all manual fire alarm pull stations accessible and unobstructed?
  - ☐ Are all fire/smoke doors accessible and unobstructed?
  - ☐ Is the digital communicator to the fire alarm monitoring center displaying "NORMAL"?
  - ☐ Where required, is an operator on duty at all times at the fire alarm monitoring center?
  - ☐ Are Interim Life Safety Measures available for implementation in the event of a fire alarm system failure or other necessitating condition/circumstance?
- Fire extinguishers:
  - ☐ Are portable fire extinguishers in their place and of the proper type?
  - ☐ Are all extinguishers easily accessible?
  - ☐ Are monthly inspections being conducted? Note: Electric space heaters may not be permitted. If this is the case, check to see if they are being used.
  - ☐ Was the last annual inspection within the past 12-months?
  - ☐ Was the last hydrostatic test within the last 5-years?
  - ☐ Are seals and tamper pins in place?
  - ☐ Are fire extinguishers free of damage, corrosion, leakage, and clogged nozzles?
  - ☐ Do pressure gauges indicate that extinguishers are ready for use?
  - ☐ Are all new staff members trained on extinguisher use?
  - ☐ Do all staff members know where extinguishers are located?
- Fire suppression systems:
  - ☐ Are all fire hydrants visible and readily accessible?
  - ☐ Do fire hydrants have caps?
  - ☐ Are fire hydrants, sprinkler systems, and Fire Department Connections (FDC) visible, accessible, and operable?
  - ☐ Are - fire department connections (FDC) visible and accessible?
  - ☐ Are fire sprinkler heads and piping in good condition?
  - ☐ Is a zone of ≥18-inches below the sprinkler head free of obstruction/materials?

- ☐ Is there a stock of sprinkler heads available in each building?
- ☐ Have the sprinkler systems been inspected by an outside contractor within the last 12 months?
- ☐ Have quarterly inspections of sprinkler systems been documented?
- ☐ Are sprinkler heads kept dust, lint, and grease free?
- ☐ Are the sprinkler systems operational?
- ☐ Does the fire alarm control panel display "fire sprinkler trouble"?
- ☐ Are kitchen hoods equipped with automatic suppression systems?
- ☐ Have kitchen hood suppression systems been tested on a 6-months basis?
- Exits:
  - ☐ Are exits clear of storage and clutter?
  - ☐ Do all staff have egress keys for pad-locks at chained/locked gates?
  - ☐ Are EXIT lights illuminated?
  - ☐ Are stairwells and corridors clear of storage/clutter?
  - ☐ Are all exterior doors fully operable and free of obstructions?
  - ☐ Are all fire doors marked with "Fire Door Keep Closed" sign?
  - ☐ Are all fire doors kept closed or equipped with a self-closing device?
  - ☐ Are all corridors, exits, ramps, and stairs illuminated to at least 1-foot candle[37]?
  - ☐ Are aisles at least 36-inches wide and clear and unobstructed?
  - ☐ Are panic bars, closers, and other hardware operating as intended?
  - ☐ Are fire doors chocked open?
- Keys:
  - ☐ Do staff members carry all necessary keys to operate the fire alarm control panels/system?
  - ☐ Do staff members carry appropriate keys to access circuit breaker panels?
  - ☐ Do shift supervisors have access to grand master key in order to access all locksets in the facility?
  - ☐ Do housekeepers carry appropriate keys to access circuit breaker panels and have they been trained to shut off a breaker switch?
- Emergency generator/fire pump:
  - ☐ Have the emergency generators been tested "under load" within the past 30-days?
  - ☐ Does each emergency generator have at least a 24-hour fuel supply?
  - ☐ Are the emergency generators maintained on a routine basis? Note: A maintenance logbook should be maintained.
  - ☐ Does the facility have a motor-driven fire pump to provide back-up pressure for the sprinkler system?
  - ☐ Has the pump been maintained and tested regularly?
- Emergency plans: See Earthquake and Other Disaster Preparedness, below.
- Fire drills:
  - ☐ Has at least one drill been conducted per shift per quarter for the past 12 months?

---

[37] Foot-candle: A unit of measure of the intensity of light falling on a surface, equal to one lumen per square foot and originally defined with reference to a standardized candle burning at one foot from a given surface.

      ☐ Was an evacuation drill conducted in the last 12-months or as required by local fire officials?

      ☐ Did staff members demonstrate quick and complete control during the evacuation drill?

      ☐ Was the room/area cleared and a head count conducted?

      ☐ Did drill simulate realistic situation/condition?

      ☐ Did the fire alarm system function properly?

      ☐ Are all drills performed at unexpected times?

      ☐ Is staff routinely trained to dial emergency phone numbers to report any/all alarms?

- Compartmentalization:
  - ☐ Have areas of refuge been identified?
  - ☐ Have staff been drilled on "evacuation in place" using the building's smoke compartmentalization?
  - ☐ Are 1-hour rated (or to code) doors or automatic sprinklers installed in boiler rooms, janitorial closets, maintenance shops, laundry rooms, kitchens, general storage rooms, and other places that you may identify?
  - ☐ Are there any firewall penetrations that are not sealed?
  - ☐ Are all fire/smoke doors fully operational?
  - ☐ Are all laundry room lint traps cleaned on a daily basis?
  - ☐ Do the laundry chute doors latch?
- Furnishings/decorations and storage:
  - ☐ Are all furnishing and decorations flame retardant?
  - ☐ Does artwork/posters or decorations cover ≤20% of open wall space in any given area?
  - ☐ Are combustibles stored properly in appropriate areas?
  - ☐ Is combustible personal property stored in non-combustible containers?
  - ☐ Are all facility wastebaskets non-flammable?
  - ☐ Are curtains, mattresses, carpets, and cushioned furniture flame retardant?
  - ☐ Are staff members who are responsible for purchasing the above items aware of the Furnishings/Decorations Procurement Policy? Note: A Furnishings/Decorations Procurement Policy that requires the purchasing of fire retardant furnishings/decorations should be in place.
  - ☐ Are combustible decorations flame retardant treated?
  - ☐ Are combustible window coverings/cubicle materials flame retardant treated?
  - ☐ Are designated staff smoking areas clean and free of potential fire hazards?
  - ☐ Are staff complying with facility smoking policy?
  - ☐ Are residents complying with facility smoking policy?
- Exterior/interior housekeeping:
  - ☐ Do all exits, emergency exits, and fire escapes afford an unobstructed passage to a safe area?
  - ☐ Are the grounds that surround the facility clear of accumulations of combustible material and brush? Note: This is especially important in fire areas such as Southern California. Check to ensure that the facility is in compliance with local ordinances.
  - ☐ Is there a facility plan to keep stored combustibles organized and to a minimum?

- ☐ Are all trash containers emptied and cleaned on a daily basis?
- ☐ Are alcoves adjacent to main corridors free from combustible materials (boxes, trash, etc.) at all times?
- ☐ Are areas under stairwells free of clutter? Note: There may be ordinances that prohibit any storage under stairwells.
- ☐ Are special areas such as shops/laboratories, eating areas, and storage and packing areas cleared of debris daily and clean? Note: Check for signs of vermin.
- ☐ Are safe capacity signs posted in assembly areas such as auditoriums and eating areas?

EARTHQUAKE AND OTHER DISASTER PREPAREDNESS

In many areas, earthquakes are not a matter of "if," they are a matter of when. For example, based on a long-term analysis of seismic activity between Northern California and Southern British Columbia, we know that the chance of a 9.0 or stronger earthquake is likely to occur sooner, rather than later in this region.[38] The recent Tohoku earthquake that struck 80 miles off of Sendai, Japan,[39] is yet another example of just how devastating earthquakes and their frequently accompanying tsunamis can be. Similarly, other disasters such as hurricanes, tornadoes, and winter storms are over the horizon; we need to prepare for them. What follows in this section uses the word earthquake, but is intended for all major "black swan" events.

- Geological and infrastructure risks:
  - ☐ Do geological hazards exist, for example, earthquakes, landslides, avalanches, flash floods, forest fires? Notes: (1) Consider also if the site is subject to inundation from a tsunami, seiche,[40] or dam or reservoir failure or flooding? (2) Is there the potential for a gas or steam line break, leakage, and explosion?(3) Is there the potential for a fire and domestic water line break, leakage, or lack of supply or pressure?(4) Is there the potential for domestic water supply contamination?(5) Is there the potential for damage due to a sewer pipe breaking?
  - ☐ Are adjacent upslope or downslope soils stable?
  - ☐ Does the site ground composition consist of fill that could increase shaking?
  - ☐ Is the ground stable or is it subject to liquefaction ground failure?
  - ☐ Are any of the following infrastructure risks present either as stand-alone risks or as the result of a geological catastrophe?
  - ☐ Are transportation, communication, and utility lifelines vulnerable to disruption? Notes: (1) Where possible, alternate routes for people, communications, and utilities should be identified. (2) Consider the affects from damaged highways, bridges, harbors, airport, rail lines; nearby power plant damage and nuclear contamination; power grid and substation

---

[38] A major earthquake in this region could have a crippling effect on major shipping ports including Vancouver, Seattle, Portland, Oakland, and Long Beach, California.

[39] March 11, 2011.

[40] Seiche: An oscillation of the surface of a pool, lake, or landlocked sea that varies in period from a few minutes to several hours.

damage; and telecommunication system damage; don't assume that wireless communications will survive a major disaster, for example, cell phones were out following the destruction of the World Trade Center and during the 2003 area-wide blackout in the Northeast.

☐ Are adjacent land uses potentially hazardous?

☐ Are hazardous materials stored or used in close proximity to buildings?

☐ Are building setbacks adequate to prevent pounding from adjacent buildings during an earthquake?

☐ Are nearby areas better left under developed to reduce earthquake risk from landslides, high liquefaction, flooding, or intensive shaking or separation from other buildings? Note: This would obviously be important to any potential buyer who may want to further develop the site.

☐ Does the site have the potential for added safety from earthquake-induced landslides by cutting unstable slopes, increasing the surface runoff, or increasing the soil water content? Note: This too would obviously be important to any potential buyer who may want to further develop the site.

☐ Is adequate space on the site for safe and defensible areas of refuge from hazard?

☐ Is there the potential for a water or sewage treatment plant shutdown and damage?

☐ Are major facilities in the proximity such as an oil refinery, major petroleum pipeline, nuclear plant, and so forth?

- Property risks: When it comes to protecting structures from an earthquake or other disaster, there is a wide range of risk mitigation techniques to consider and adopt. Experienced professionals, for example, architects and structural and soils engineers can assess the level of property and related building earthquake risk. A risk engineer can work with these professionals to develop an overall risk mitigation plan.

☐ If the buildings that you are inspecting predates current codes, has an effort been made to adequately retrofit them for an earthquake?

☐ Does the buildings have any construction quality issues that may make them vulnerable to earthquake damage? Note: Look for recent nonstructural, structural, or settlement damage?

☐ Could the buildings survive the largest credible earthquake for the area and be functional afterward? Note: A structural engineer may have to answer this question.

☐ Have the fire sprinkler systems been designed to for an earthquake?

☐ Are emergency response and protection plans in place? Note: Plans should include actions to take in case of fire sprinkler leakage; gas, water supply, waste water, and other process pipe rupture; structural damage; nonstructural damage; fire and explosions; power loss; security after an earthquake in place?

☐ Are your utility feeds into the building designed to handle an earthquake?

☐ Are seismic gas valves for utilities and monitoring systems for critical operations in place?

☐ Have large pieces of equipment been secured?

☐ Have heavy furnitures such as bookshelves and file cabinets been strapped or attached to walls?

□ Are areas under desks and tables uncluttered? (So people can take cover.)

□ Are doors, exits, and aisles clear for quick evacuation?

□ Are all fire extinguishers charged? Are they inspected monthly?

□ Are fire alarms, sprinklers, and emergency lighting scheduled for testing? Are logs kept of all tests?

□ Are flashlights, a battery-operated radio, extra batteries, and a first aid kit available to the facilities staff and other designated people? Note: These items should be checked regularly.

□ Are evacuation routes inspected to make certain they are not cluttered? Note: Check to ensure that these routes will not be blocked by fallen debris.

□ Is the local fire department familiar with the facility's emergency plan and any recent changes?

□ Has the local fire department visited the facility and reviewed emergency plans within the past 12 months?

- Operational risks: Since operational risks are not physical plant related, they are generally not included in due diligence/capital plan program inspections. Nevertheless, you may wish to add your two-cents to some of the following issues if you have opinions about them:

  □ Is a plan in place to deal with the complete shutdown of production due to a disaster? Note: some issues to consider are: loss of key employees (is a succession plan in place?); fatalities and injuries of employees; key employees not available at the time of the occurrence; equipment replacement (is there a inventory of all key equipment?); site becoming inaccessible; the loss of communications and utilities; the likelihood of environmental contamination; supply chain disruption; security risks following a disaster; a list of emergency contact phone numbers for all employees and management (the list should include family contacts, schools, doctors, and medical needs.)[41]; a damage survey to determine the need for temporary relocation or timing of re-occupancy; and clean up of company properties including securing of contractors to support and supplement crews in the repair of damaged facilities

  □ Have mitigation plans that address emergency needs, including loss of utilities, and potential business losses and recovery been developed?

  □ Are key suppliers aware of these plans and have they developed back-up plans for the organization?

  □ Has a value chain that assesses the potential affect and extra recovery expenses of a disruption in supplies been developed?

  □ Does the organization review emergency procedures regularly?

  □ Are drop-cover-and hold drills conducted regularly? Note: People should be designated to command co-workers to DROP—COVER—HOLD ON and stay in that position until the ground stops moving.

---

[41] Employees should be advised to have the following information: Medicare and other health insurance cards; copies of family members' names, addresses, phone numbers, etc.; a list of primary care managers, other doctors and phone numbers; emergency contact names and phone numbers; known prescription medications and doses; a list of allergies; style model and serial numbers for any medical devices; extra batteries for wheelchairs and hearing aids; and personal items such as eyeglasses and other special equipment. Arrangements should be made for childcare.

- ☐ Do people know the safest place in their immediate area?
- ☐ Are there people trained in first aid and cardiopulmonary resuscitation (CPR)[42] skills? Note: Daily occupants should know where first aid items are located.
- ☐ Are emergency supplies such as sanitary supplies and toilet paper on hand?
- ☐ Has an emergency evacuation plan that details how employees will evacuate from the building, where they will meet, how to account for everyone, and how to get further instructions been developed? Notes: (1) Emergency plans should designate responsibilities and disaster buddies for people with physical disabilities. (2) Emergency plans should identify people who have specialized knowledge such as knowing how to shut off utilities. (3) Responsible people such as floor managers and floor wardens should be prepared to:
  - – Check the evacuation routes.
  - – Assist co-workers to evacuate to predesignated assembly points.
  - – Account for all employees.
  - – Give first aid where appropriate.
  - – Call 911 for life-threatening emergencies.
  - – Conduct a building survey and check utilities.
  - – Tune into the portable radio for information and instructions.
  - – Decide whether employees should re-enter the facility or go home to check on their families (if roads are open).
  - – Determine if the facility can be reopened. Notes: (1) Clean up teams should be formed to clear debris, clean up spills, and check for other hazards. (2) Plans should include the securing of contractors to supplement in-house.
  - – Remind co-workers of possible after-shocks and to be prepared to once again follow procedures.
  - – Hold an employee meeting to discuss fears and emotions.
  - – Provide employees with needed assistance.
  - – Contact key suppliers to resume production and deliveries or to activate earlier developed back-up plans. Note: Alternate sources of essential supplies and replacement parts should be identified should normal vendors be unable to function after a disaster.
  - – Are plans and lists reviewed, tested, and updated at least quarterly?
  - – Do people know to check themselves and the people around them for injuries?
- ☐ Have plans been established to conduct a damage survey of the facilities to determine the need for temporary relocation or timing of re-occupancy?
- ☐ Have long-term plans been developed to provide for re-establishing the full functioning of the company?

---

[42] Cardiopulmonary resuscitation (CPR): A procedure to support and maintain breathing and circulation for a person who has stopped breathing (respiratory arrest) and/or whose heart has stopped (cardiac arrest). CPR is performed to restore and maintain breathing and circulation and to provide oxygen and blood flow to the heart, brain, and other vital organs. While trained bystanders and healthcare professionals can perform CPR on infants, children, and adults, contractors should consider having someone on each of their jobs trained in CPR; it should always be performed by the person on the scene who is most experienced in CPR.

VERTICAL TRANSPORTATION

There is no need to stress the importance of elevator safety. In truth elevators are quite safe, but when an accident occurs, they are often very serious, causing great harm or death. For many people they are a source of fright. And, if you have ever experienced being stuck in a crowded elevator for an extended period, you fully realize the importance of a well-functioning elevator. Generally a licensed elevator inspector would conduct an elevator due diligence/capital project planning inspections. See: High-Rise Building Construction, Conveying Systems and Final Inspections—High Rise Building Construction, Elevators.

Appendix A
INSPECTING
A SCHOOL DISTRICT,
A CAMPUS, OR
SIMILAR[1]

I shall risk life and limb and briefly wade into the hot potato of inspecting and doing capital planning for a major city's school buildings,[2] and by extension police stations or firehouses. All cities do this in one form or another, but since I had some minimum experience with New York City's School Construction Authority's (SCA) program, that I thought was excellent, I'll pass this experience on.

Since we are talking about public buildings that are spread across a city or county, it is important that the inspection program be equitable, that is, that it not skew the results in favor of one school or district and that the resulting capital plan optimizes the distribution of scarce resources. To accomplish this, the inspection program must inspect all buildings based on the same criteria despite the inspections probably being conducted by different people. In the case of the SCA, it had been mandated to develop a five-year capital program after several serious accidents had occurred as a result of facility failures.

The agency that is responsible for the maintenance, repair, and alterations of the buildings must decide just how it will use the information that inspectors will gather. Will it be to develop a capital plan, a maintenance program, or both? This determines the extent of the inspections. However, one thing is certain: the inspection reports

---

[1] School buildings are used in the discussion, but the program is valid for other buildings too.
[2] A quick check on the web shows that New York City has approximately 1,700 schools, Los Angeles about 1,100 schools, and Chicago about 700 schools.

and any ensuing plan that results from them will be used to obtain funding, be it from elected officials or top-level management.

If the agency has not performed a program of periodic inspections, a good place to begin is a thorough review of the files. There is probably a considerable amount of information about each building in them, but as a minimum such things as building address, size of building, year constructed, year of last major renovation (roof replacement, window replacement, etc.) should be available. If the opportunity presents itself speaking with key staff members at each school is another good source of information. This information will go on the heading of each building's inspection. It will be helpful in assessing the inspection results. For example, people may find following completion of the inspection program that as a group all 50-year-old buildings will require major renovations, but those that are 70 years old but were renovated 20 years ago are in fairly good shape. This kind of information helps when prioritizing, grouping contracts to gain economy of scale, and seeking funding.

The agency needs to have a handle on just how much it can afford on gathering information and how it will accomplish them; the SCA decided to have an outside contractor conduct the inspections (more on this later). This will determine how detailed the inspections will/can be. There are many variants to how long it will take to completely inspect a school building for capital planning, for example, size, age, quality of the original construction, and prior maintenance. However, from experience a team of one architect, one electrical engineer, and one mechanical engineer (one of the team being licensed, the others degreed or very experienced and one appointed as the team leader) each having a well thought-out checklist that is loaded into a computer tablet completed a fairly typical school building in around four hours. You may want to add one person to the team if the grounds are extensive or you are aware of other issues.

Of course these teams need to be equipped: computer tablets, inspection tools (tape measures, telephones/cameras, flashlights, log books, etc.), and a team vehicle. In the case of New York City, obtaining vehicles was not as easy as it sounds, as there were many teams in the field for a relatively short time; fleet rentals are usually for a longer period and not every rental company wanted this business.

Staffing also included a disciple chief for each discipline. These were senior people. They had an extensive library of photos and they gave classes so that the inspectors would know what to anticipate. The chiefs also took the entire group to several schools where the new inspectors conducted their first inspections and had them graded. The chiefs remained with the project until the end. They reviewed all reports (they were computerized) and continued to evaluate and instruct. There was also one team that performed follow-up quality assurance inspections.

A few words about checklists: In the case of the SCA, these were developed jointly between the contractor and the SCA; they were for architecture, mechanical, and electrical. Because a team would move from one school to another and each school was of a different age and style, the checklists had to be valid for all. Using the mechanical checklist as an example, the list would show all possible boilers, one after the other, and the inspector would have to check off all none applicable boilers. As he/she moved down the checklist, each line had to receive an answer, that is, satisfactory, not satisfactory, or not applicable. Then for the applicable boiler, the checklist would carry the inspector into the details of the particular boiler.

The checklists established the level of detail of the total survey. This was based on the SCA's needs. Some checklist items required a relatively quick look at an item and a quick condition rating for each component. For other items, the checklist guided the inspector to delve more deeply. Deficiencies were identified with appropriate units of measure (e.g., square feet of pointing, numbers of windows needing replacement, etc.); where appropriate photos and sketches were inserted into the reports. Some checklist items required inventorying equipment and adding data plate information of major items. ADA assessments were part of all checklists.

It would be wonderful if your inspectors could also assess the overall adequacy of each building to meet current needs, as this could cause a decision to replace a building even if it is in good shape. But this last item probably exceeds the ability of most checklists and inspectors and will have to be done separately by people who are educators not building inspectors.

Some nitty-gritty information:

- All inspectors would gather early Monday mornings where assignments were handed out and problems and lessons learned were discussed. Tablets were handed out.
- Each team had a car. One man would take it home at night. He/she would arrange to pick up the other members each day. Since New York City has a great public transportation system, this was usually at a subway station.
- Teams were allowed to go home early, but only after all disciplines were finished; no team was ever found to be cheating. Teams called the main office at the start and finish of each day.
- Inspection reports were stored in the tablets until Friday evening, at which time they were dropped off where the reports were uploaded. These reports went to the client, the SCA, and the discipline chiefs. The chiefs reviewed reports during the week and followed up on potential errors, for example, 20,000 square feet versus 2,000 square feet. The SCA did not receive written reports. Square feet is OK, better in fact.
- Office staff called schools to be inspected one week in advance. Team leaders called the schools the day prior to their arrival. On arrival, the team usually met briefly with the principal or vice-principal.
- Inspections were scheduled beginning at the outskirts of the city. As the circle got smaller, people were let go and teams were sometimes reformed based on management's evaluation of performance.
- Because expenses went to a public agency telephone bills, car usage, and so forth, were carefully monitored and audited.
- Throughout there was a close working relationship between the SCA and contractor. Occasionally, SCA staff members visited inspection teams.

Given the large number of buildings involved and the scarcity of funds, the SCA had to prioritize findings. The SCA broke down inspection results into five categories:[3] (1) Good condition (the system/subsystem is sound and performing well); (2) condition

---

[3] The SCA's inspection program is presented in some detail because, besides having some firsthand knowledge of it, I believe the program is a very sophisticated program that includes well thought-out inspections along with a logical and very transparent decision process.

between good and fair; (3) fair condition (the component is performing adequately, but may require preventive maintenance to prevent further deterioration and restore it to good condition); (4) condition between fair and poor; and poor condition. (The component cannot continue to perform its original function without repairs or is in such a condition that failure is imminent. Equipment that exceeded its useful life and required replacement would also fall under this category.)

The SCA then attached a recommended action to each deficiency and then a "Purpose of Action" to every recommended action. The following menu is taken from the SCA model and is listed in priority: (1) life safety, (2) structural, (3) regulation/code, (4) security, (5) betterment, (6) cost avoidance, (7) operations and maintenance savings, (8) esthetics, and (9) community. As an example, using this coding system, a fan coil unit that is beyond its useful life (poor condition) but in a basement storage area would probably be deemed a betterment, while loose bricks in a façade (poor condition) would be deemed life safety and, thus, take precedence.

The SCA also looked at urgency, by attaching five ratings: (1) fail now, (2) fail within 6 months, (3) fail within 24 months, (4) no fail within 24 months, and no urgency.

After all the components were inspected and rated, weights were assigned to the various systems and their components. And then, using these assigned weights they used a condition roll-up system that gave management those buildings/systems/components that are critical or of the highest priority. It is from this that the SCA developed its 5-year capital plan.

It has been some time since I went back to see how things are now going, but in the past the SCA went through the process just described every 5 years. It also followed up with a far less comprehensive inspection program annually.

Inspection Reports: By now checklists have probably worn you out, so I shall present only abbreviated versions of the checklists that you might develop for your inspection teams; your checklists should be tailored to your situation. Since inspectors will be inspecting many buildings, old and new, your checklist should include all the items and situations possible. To ensure that the inspector covers everything, each question needs to be answered, even if the answer is not applicable (NA). These checklists would be loaded into a computer tablet that an inspector would carry. The inspector would enter information as he/she proceeds; photos and sketches would be added as necessary. Again, the inspector must answer each question before moving on, even if it is an NA. Below I attempt to provide items that you may want an inspector to survey. But, when all is said and done, you will have to create the checklists to meet your needs.

- **Architectural—General:** Each checklist, which eventually becomes an inspection report, should contain the following building data: Area of entire property, area of building, number of stories and whether there are basement levels; year built; number of people on the facilities staff; installed system, for example, cable TV; and year various building components were upgraded (e.g., roofing including flashing, coping stones, inner wythe brick: 2010. Third floor windows replaced: 2008.).

  Inspectors should record the key people and staff personnel with whom they come in contact, the general condition of the facility, and anything out of the ordinary, for example, no gas service to building.

ARCHITECTURAL:

- **Asset data:**
- **Hazardous conditions:**

Question	Answer

- **Accessibility:**
  - Exterior routes
    - Curb cuts and turns
      - Compliance? — Complies.
    - Parking
      - Exists? — Yes
      - Compliance? — Deficiency. No accessible stalls.
    - Lifts
      - Exists? — No.
      - Required? — Yes.
    - Ramps and railings
      - Exists? — Yes.
      - Compliance? — Complies.
    - Stairs and railings
      - Exists? — Yes.
      - Compliance? — Complies.
    - Entrances and exits
      - Compliance? — Deficiency.
      - Handles, pulls, and so forth have
      - A shape that is easy to grasp? — Complies.
      - Saddle height ≤½ inch? — Saddle height >½ inch.
  - Interior routes
    - Corridors and lobbies
      - Compliance? — Deficiency.
        - Change in elevation? — Change in elevation.
    - Lobby doors and hardware
      - Compliance? — Complies
    - Corridor doors and hardware
      - Exists? — No.
      - Required? — No.
    - Drinking fountains
      - Exists? — Yes
        - Control not on front? — Control not front mounted.
        - Spout height ≤36 inches? — Complies. Spout height ≤36 inches.
    - Ramps, stairs, and railings
      - Exists? — Yes.
      - Compliance? — Complies.
    - Elevators
      - Exists — No.
    - Rooms and spaces
      - Classrooms
        - Exists? — Yes
        - Compliance? — Complies.

And so it would go. You can build your own checklist based on what you know about your buildings. There is a great deal of information regarding ADA Accessibility Guidelines (ADAAG), among them United States Access Board, *ADA Accessibility Guidelines (ADAAG)* and local building codes.

Other major headings that you might consider for your architectural inspections (of schools):

- **Structural engineer required?**   Space for notes
- **Exterior:**

Areaway walls	Areaway slab
Areaway gratings	Awnings and Canopies
Chimneys	Coping
Façades	Cornices
Doors	Lintels
Transom/side lights	Exterior walls
Loading dock	Louvers
Parapet walls	Roofing
Bulkhead walls	Cupolas
Dormers	Flashing?
Hatches	Roof barriers
Ladders	Skylights
Leaders, gutters, downspouts	Roof drains
Exterior stairs/ramps	Exterior walls
Railings	Stairs/ramps
Windows	

- **Interior:**

Classrooms/corridors/administrative spaces	Doors 7 door hardware
Floors	Walls
Gymnasium (break down to include such things	Interior window guards
as ceiling, fixed equipment, flooring,	Locker rooms and showers
score board, and so forth	Pools
Specialties, for example, cabinet work	Chalk/white boards
Structural	Stairs: Here is an example of how your computerized checklist may look:

Question	**Answer**
Stairs: Interior	
Concrete	
Exists?	No
Condition?	NA
Deficiencies?	NA
Deficiency type?	NA
Deficiency location?	NA
Deficiency quantity?	NA
Deficiency unit of measure?	NA
Potential action?	NA

Urgency of Action?	NA
Purpose of Action?	NA
Photo?	NA
Violations?	NA

Metal

Exists?	Yes
Condition?	4 – between fair and poor
Deficiencies?	Yes
Deficiency type?	Rust/deterioration
Deficiency location?	All stairwells
Deficiency quantity?	200
Deficiency unit of measure?	SF
Potential action?	Repair
Urgency of Action?	4 – Address within 6 months
Purpose of Action?	2 – Structural
Photo?	No
Violations?	No

Stone

Exists?	Yes
Condition?	3 – Fair
Deficiencies?	No
Deficiency type?	NA
Deficiency location?	NA
Deficiency quantity?	NA
Deficiency unit of measure?	NA
Potential action?	NA
Urgency of Action?	NA
Purpose of Action?	NA
Photo?	NA
Violations?	NA

Handrails - metal

Exists?	Yes
Condition?	3 – Fair
Deficiencies?	No

Handrails—wood

Exists?	No
Condition?	NA
Deficiencies?	NA

Partitions

Exists?	No
Condition?	NA
Deficiencies?	NA

- **Life Safety:**

Fire escapes   Holding areas

- **Site:**

Playground Irrigation system
Water fountains   Drainage system
Fences   Site walls (other than

Paving (Non-vehicular)   retaining walls)
Paving (Vehicular area)   Retaining walls
Sidewalks   Site structures (flag poles, bulletin
                                  boards)

MECHANICAL:

- **Asset data:**
- **Hazardous conditions:**
- **Acid waste/drain and vent:**
- **Air conditioning:**
  Chiller water piping   Chiller, centrifugal water-cooled
  Chiller, absorption   Chiller, reciprocating air-cooled
  Chiller, scroll/rotary air-cooled   Chiller, reciprocating water-cooled
  Chillers, emergency stop switch Chiller, scroll/rotary water-cooled
  Cooling tower   Cooling tower pump
  Air cooled condenser   Dx air-cooled condensing unit
  Fan, return   Package/rooftop unit
  Split heat pumps   Window type AC
- **Climate control:**
  Temperature control air compressor   Temperature control subpanel DDC[4]
  Temperature control pneumatic   Temperature control pneumatic
  valve   subpanel Ddc
  Temperature control refrigerated   Temperature control steam trap
      air dryer
  Direct digital control system (DDC)   Pneumatic tubing

Temperature control thermostat. Here is
an example of how your computerized
checklist may look:

Question	Answer
Temperature control thermostat	
Exists?	Yes
Condition?	4 – Between fair and good
Deficiencies?	Yes
Deficiency type?	Deteriorated
Deficiency location?	Building
Deficiency quantity?	30
Deficiency unit of measure?	Each
Potential action?	Replace
Urgency of Action?	4 – Address within 6 months
Purpose of Action?	5 – Restore
Photo?	No
Violations?	No
Temperature control: pneumatic pressure reducing valve	

---

[4] Direct Digital Control (DDC)

- **Compactor:**
- **Conveying equipment:**

Dumbwaiters	Elevators
Escalators	Sidewalk ash hoist

- **Domestic water system:**

Electric pressure booster system	Pressure reducing valves
Heat exchanger	Water heater
Hydraulic/pneumatic booster system	piping
Backflow preventer assembly	

- **Drain/waste/vent and storm system:**

Floor drains	Piping
Storm system	

- **Emergency generator:**
- **Fixtures:**

Lavatories	Sinks
Toilets	Urinals
Showers	Emergency showers
Eye washes	Sink and fountain combos
Drinking fountains	

- **Gas service:**

Compressor booster	High-pressure distribution piping
Low-pressure distribution piping	Meter
Piping (indoor)	Science lab service

- **Heating:**

Expansion tank	Heating surface – cast iron radiator
Heating surface – commercial type convectors	Heating surface – fin tube
Heating surface – heating coils	Hot water heat exchanger
Pressure reducing valve (LPS)	Steam & condensate piping

- **Heating plant:**

Boiler auxiliaries	Boiler emergency stop switch
Boiler feed water system	Boiler feed water treatment
Boiler leader piping/valve	Boiler safety valves
Piping, waterline (makeup)	Steam condensate return gravity system
Steam condensate return vacuum system	
Motors/motor starters	Boiler system, for example, steam, hot water
Gas fired furnace	Fuel system, for example, gas, oil, panel, combustion air system
control Gas meter room exhaust fan/vent	
Oil leak detection/gas leak detection	Overfill alarm
Fuel oil level gauge	Fuel oil piping
Fuel oil pump heater set	Fuel tank heater
Fuel transfer pumps	Fuel transfer pump motors/starters
Gas train and vents	

- **Kitchens:**

Range hood exhaust fans	Range hood ductwork

Range hood fire suppression system   Gas system
- **Piping:**
Cold water (HVAC makeup)   Hot water hydronic[5] supply
Hot water hydronic return   Steam
Steam condensate return
- **Swimming pool:**
Filters   Heater
Pumps
- **Pumps:**
Ejectors (sewage)   Chilled water pumps
Hot water circulating pumps   Chilled water pump motors/starters
- **Sprinklers, fire system:**
Booster pumps   Booster pump motors/starters
Water gong[6]   Deluge tank/piping
Piping   Heads
- **Toilet rooms:**
Ceiling   Floor
Stalls   Walls
Fixtures: See Fixtures
- **Ventilation:**
Science lab hood exhaust fan   Air terminal box
Fire damper   Smoke damper
Fire, smoke damper (FSD)   Central station air handler
    combination
Ducts/registers/ductwork   fan coil units
Exhaust fans   Supply fans
Unit ventilators

ELECTRICAL:

- **Asset data:**
- **Hazardous conditions:**
- **Functional areas:**
Auditoriums   Computer labs
Kitchens   Libraries
Science labs
- **Local sound system:**
- **Bell systems:**
- **Emergency DC standby battery power:**
Lead-acid battery bank   Nickel-cadmium battery bank
- **Emergency generator set:**
Emergency lighting   Exit lights

---

[5] Of or relating to a heating or cooling system that transfers heat by circulating a fluid through a closed system of pipes. Hydronic systems usually use a pump to circulate the liquid.

[6] A hydraulically operated outdoor alarm that is used with fire protection systems. Gongs are used in conjunction with alarm check, dry pipe, deluge, and preaction valves to sound a local alarm.

- **Emergency power:**

Automatic transfer switches	Battery chargers
Relay test switches	

- **Exit lights:**

Battery packs	24-hour panel

- **Fire alarm system:**

Bells/horns	Bells at panel
Duct smoke detector – ionization	Elevator recall
City/county fire alarm system	Fire alarm control panel (electronic)

Fire alarm control panel (standard). Here is an example of how your computerized checklist may look:

Question	Answer
Exists?	Yes
Condition?	3 – Fair
Specifics	
Control panel location	Boiler room
Manufacturer	Pyrotronics
Equipment ID	FIRE ALARM
Capacity/size	32 zones
Installation date	11/15/65
Deficiencies	No
Fire alarm strobe	Fire pump
Fused cutout panel	Magnetic door holder contactors
Manual pull station	Photoelectric smoke detector
Sprinkler flow switch	Tamper flow switch

- **Grounding system:**
- **Intercom system:**
- **Lighting**

Exterior security	Stage, theatre (incl. dimming)
Fluorescent (Pendant, recessed, surface)	Incandescent (Pendant, recessed, surface)

- **Lightning protection:**
- **Motor control center:**

Combination – circuit breaker type	Combination – fused type

- **Motor starter/contactor:**

Combination – circuit breaker type	Combination – fused type

- **Panelboard:**

Fused disconnect switches	Fused knife switches
Fused toggle switches	Molded case circuit breakers

- **Public address systems:**
- **Security:**

Central control panel/Intrusion alarm	Closed circuit television system
Intrusion alarm – infrared	Intrusion alarm – ultrasonic

- **Switchboard:**
  Air circuit breaker   Fused knife switch
  Fused disconnect switch.   Molded case circuit breaker
- **Telephone:**
  PBX/Intercom   Standard
- **Transformer:**
  Dry type   Liquid type
- **TV cable system:**
  Cable service   Master antenna system

Appendix B
ORGANIZATIONS THAT AFFECT CONSTRUCTION

(Extracted from Schmid, K.F. *Concise Encyclopedia of Construction Terms and Phrases.* New York, NY: Momentum Press, LLC.)

Contractors (and inspectors) are called upon to build (inspect) many things, and not all are straightforward. Below are listed organizations that you may wish to contact when that unusual project comes your way or, perhaps, merely to broaden your knowledge base. You may also want to join one or more of them for purposes of networking, information, continuing education, and gaining the benefit of their advocacy.

Acoustical Society of America (ASA): The society provides information in the broad field of acoustics.

Adhesive and Sealant Council (ASC): The council provides information, education, and representation to its members.

Air Barrier Association of America (ABAA): A resource for air barrier education and technical information.

Air Conditioning Contractors of America (ACCA): A trade association that promotes professional contracting, energy efficiency, and healthy, comfortable indoor environments.

Air-Conditioning, Heating and Refrigeration Institute (AHRI): Member companies produce residential and commercial air conditioning, heating, and water heating equipment. The institute provides members with a certification program, standards, advocacy, and other activities.

Air Diffusion Council (ADC): The council represents manufacturers of flexible airducts and air connectors.

Air Infiltration and Ventilation Centre (AIVC): An international organization that provides industry and research organizations with technical support aimed at optimizing ventilation technology.

Air Movement and Control Association International (AMCA): An international association of air equipment manufacturers.

Alliance for Fire and Smoke Containment and Control (AFSCC): An alliance of building enforcement, construction, design, and manufacturing professionals. It promotes a balanced fire protection design in the built environment.

Allied Board of Trade, Inc. (ABT): ABT acts as a liaison between the designer and supplier to promote ethical standards of business conduct in the industry. Members have access to trade information, website design, magazine program, source information, and business advice.

Aluminum Anodizers Council (AAC): Members are engaged in aluminum anodizing, are suppliers of products and services used in the anodizing of aluminum products, or are purchasers of anodized finishes.

Aluminum Association: An international trade association that promotes aluminum as a sustainable and recyclable automotive, packaging, and construction material. The association represents United States and foreign-based primary producers of aluminum, aluminum recyclers, and producers of fabricated products, as well as industry suppliers.

Aluminum Extruders Council (AEC): An international trade association that is dedicated to advancing the effective use of aluminum extrusion in North America.

American Arbitration Association® (AAA): The association (AAA) has a long history providing alternative dispute resolution to individuals and organizations that strive to resolve conflicts out of court. Services include assisting in the appointment of mediators and arbitrators, setting hearings, and providing information on dispute resolution options, including settlement through mediation. Additionally, AAA designs and develops alternative dispute resolution (ADR) systems for corporations, unions, government agencies, law firms, and the courts. Abroad, these services are provided through AAA's International Centre for Dispute Resolution® (ICDR).

American Architectural Manufacturer's Association (AAMA): A national trade association that establishes voluntary standards for the window, door, and skylight industry. AAMA comprises the following organizations: American National Standards (ANSI); American Architectural Manufacturers Association (AAMA); and National Wood Window and Door Association (NWWDA).

American Association of Automatic Door Manufacturers (AAADM): A trade association of power-operated automatic door manufacturers.

American Association of State Highway and Transportation Officials (AASHTO): A regulatory organization that governs the design and specifications of highway bridges.

American Backflow Prevention Association (ABPA): An organization dedicated to education and technical assistance that is focused on protecting drinking water from contamination through cross-connections.

American Bio-Recovery Association (ABRA): A trade organization that provides an international program to test and certify students who have completed ABRA-approved training courses. Courses include, among others, crime scene cleaning and approaching a horrific and graphic scene in order to clean and sanitize it and in what order to use chemicals and equipment.

American Boiler Manufacturer's Association (ABMA): The national trade association of commercial, institutional, industrial, and electricity-generating boiler system manufacturing companies (>400,000 Btuh heat input). It is dedicated to the advancement and growth of the boiler and combustion equipment industry.

American Coal Ash Association (ACAA): A trade association that is dedicated to recycling the materials that are created when coal is burned to generate electricity.

American Coatings Association (ACA): An organization that is in support of the paint and coatings industry.

American Composites Manufacturers Association (ACMA): A trade group that represents the composites industry.

American Concrete Institute (ACI): The institute is dedicated to advancing concrete knowledge by conducting seminars, managing certification programs, and publishing technical documents.

American Concrete Pavement Association (ACPA): An organization comprising concrete paving contractors, cement, and material producers, equipment manufacturers, and any company with an interest in concrete airports, highways, roads, streets, and industrial pavements.

American Concrete Pipe Association (ACPA): An advocate for the concrete pipe industry. Its members are committed to environmental improvement by producing quality concrete pipe for drainage and pollution abatement.

American Concrete Pressure Pipe Association (ACPPA): The association provides technical information and design assistance on the uses of concrete pressure pipe (CCP) to the water and wastewater industries.

American Council for Accredited Certification (ACAC): ACAC promotes awareness, education, and certification for professionals. ACAC's mission is to establish credible certification programs that provide value to its clients and the public. As part of an agreement with the Indoor Air Quality Association (IAQA) and the Indoor Environmental Standards Organization (IESO) the council agreed to focus exclusively on certification programs.

American Fence Association (AFA): The association represents the fence, deck, and railing industry in the United States and parts of Canada. AFA offers educational, certification options, and networking to its members.

American Forest and Paper Association (AF&PA): AF&PA is a trade association that advances a sustainable U.S. pulp, paper, packaging, and wood products manufacturing industry through fact-based public policy and marketplace advocacy. Member's products are made from renewable and recyclable resources and they are committed to continuous improvement through the industry's sustainability initiative: *Better Practices, Better Planet 2020.*

American Fire Sprinkler Association (AFSA): An international association that represents open-shop fire sprinkler contractors. The association promotes the use of automatic fire sprinkler systems and offers educational advancement to its members.

American Floorcovering Alliance (AFA): The alliance promotes the industry's products and services and educates its members and others through seminars, press releases, and trade shows.

American Galvanizers Association (AGA): An association that represents the post-fabrication hot-dip galvanizing industry.

American Gas Association (AGA): AGA represents companies that deliver natural gas to customers. AGA advocates the interests of its members and their customers.

American Hardboard Association (AHA): A trade organization of manufacturers of hardboard products for exterior siding, interior wall paneling, furniture, and industrial and commercial products.

American Hardware Manufacturers Association (AHMA): A trade organization with membership open to any company headquartered in the United States and is engaged in the manufacture of goods for hardware and home improvements, lawns and gardens, and painting and decorating.

American Industrial Hygiene Association (AIHA®): A nonprofit international association that serves occupational and environmental health and safety professionals practicing industrial hygiene in industry, government, labor, academic institutions, and independent organizations. AIHA administers education programs in addition to operating several laboratory accreditation programs.

American Institute of Architects (AIA): A professional organization that unites in fellowship the members of the architectural profession in the United States.

American Institute of Building Design (AIBD): AIBD is a professional organization that strives to protect and enhance its member's ability to practice their profession. The institute provides continuing education in technology, materials, and building codes. AIBD maintains professional relationships with other relevant trade, business, and professional organizations within the design and construction industry.

American Institute of Chemical Engineers (AIChE): A professional organization for chemical engineers, as distinct from chemists and mechanical engineers. There are student chapters at various universities around the world that tend to focus on providing networking opportunities in both academia and in industry.

American Institute of Steel Construction (AISC): A technical specifying and trade organization for the fabricated structural steel industry in the United States. It publishes the "Manual of Steel Construction."

American Institute of Timber Construction (AITC): AITC is a national technical trade association of the structural

glued laminated timber industry. The institute publishes the *Timber Construction Manual*, which is now in its 6th edition.

American Insurance Association (AIA): A leading property-casualty insurance trade organization. AIA represents over 300 insurers that write more than $110 billion in premiums each year.

American Iron and Steel Institute (AISI): The institute promotes the interests of the iron and steel industry.

American Lighting Association (ALA): A trade association that represents the lighting industry. Its membership includes lighting, fan and dimming control manufacturers, retail showrooms, sales representatives, and lighting designers.

American Lumber Standard Committee, Inc. (ALSC): ALSC comprises manufacturers, distributors, users, and consumers of lumber. ALSC serves as the standing committee for the American Softwood Lumber Standard (Voluntary Product Standard 20) and administers an accreditation program for grade-marking lumber produced under the system.

American National Standards Institute (ANSI): A clearinghouse organization for all types of standards and product specifications.

American Nursery and Landscape Association (ANLA): The association provides education, research, public relations, and representation services to its members. Members grow, distribute, and retail plants of all types. They also design and install landscapes for residential and commercial customers.

American Plywood Association (APA): *APA-The Engineered Wood Association* focuses on helping its industry create high-quality structural wood products. APA promotes new solutions and improved processes.

American Rainwater Catchment Systems Association (ARCSA): ARCSA promotes rainwater catchment systems in the United States. Members include professionals working in city, state, and federal government, academia, manufacturers and suppliers of rainwater harvesting equipment, consultants, and other interested individuals; membership is not limited to the United States.

American Shotcrete Association (ASA): An organization of contractors, suppliers, manufacturers, designers, engineers, owners, and others with a common interest in promoting the use of shotcrete.

American Shutter Systems Association (ASSA): ASSA provides tested and approved hurricane shutter products for its licensed members who are located throughout the coastal United States and the Caribbean.

American Society of Civil Engineers (ASCE): The oldest national professional engineering society in the United States. It is dedicated to the advancement of the individual civil engineer and the civil engineering profession through education.

American Society of Concrete Contractors (ASCC): Members include contracting firms, manufacturers, suppliers, architects, specifiers, and distributors of the concrete industry.

American Society of Furniture Designers (ASFD): The society is dedicated to advancing, improving, and supporting the profession of furniture design and its impact in the marketplace.

American Society of Heating, Refrigeration and Air Conditioning Engineers (ASHRAE): A national association that establishes standards for building energy performance.

American Society of Home Inspectors (ASHI): A professional association for home inspectors. It establishes and advocates high standards of practice and has a strict code of ethics for its members.

American Society of Interior Designers (ASID): Comprising designers, industry representatives, educators and students, AISD provides education, knowledge sharing, advocacy, community building and outreach to and for the interior design profession.

American Society of Irrigation Consultants (ASIC): The society strives to enhance the role of the independent professional irrigation consultant.

American Society of Landscape Architects (ASLA): ASLA provides education and participates in the stewardship, planning, and design of cultural and natural environments.

American Society of Mechanical Engineers (ASME): The society promotes the art, science, and practice of multidisciplinary engineering and allied sciences worldwide.

American Society of Plumbing Engineers (ASPE): An international organization for professionals involved in the design, specification, and inspection of plumbing systems.

American Society of Professional Estimators (ASPE): The society serves construction estimators by providing education, fellowship, and opportunity for professional development.

American Society of Sanitary Engineers (ASSE): Members include all disciplines of the plumbing industry.

American Society of Theatre Consultants (ASTC): The society informs owners, users, and planners about the services that theatre consultants offer for projects large and small, whether for new construction or remodeling/renovation.

American Soil and Foundation Engineers (ASFE): ASFE aids geo-professionals to achieve business excellence and to manage risk through advocacy, education, and collaboration.

American Sports Builders Association (ASBA): ASBA provides information on tennis courts, running tracks, fields, and indoor sports facilities.

American Subcontractors Association, Inc. (ASA): A trade organization that is dedicated to improving the business environment in the construction industry.

American Supply Association (ASA): An association that focuses primarily on plumbing (similar to AWWA).

American Walnut Manufacturers Association (AWMA): An international trade association that represents manufacturers of walnut lumber, dimension lumber, veneer, walnut squares, and gunstock blanks.

American Welding Institute (AWI): An organization established to bridge the gap between the findings of basic welding research and the needs of the industry.

American Welding Society (AWS): An organization whose major goal is to advance the science, technology, and application of welding and related joining disciplines.

American Wire Producers Association (AWPA): A trade association for the ferrous wire and wire products industry in North America. Members include wire producers, manufacturers and distributors of wire rod, and suppliers of machinery, dies, and equipment to the wire industry.

American Wood Preservatives Association (AWPA): The association seeks to improve the performance and longevity of sustainable wood products. AWPA is a resource for knowledge on all aspects of wood protection.

American Water Works Association (AWWA): An association for people associated with water supply (waterworks), including all types of materials.

American Wood Council (AWC): The council represents the North American traditional and engineered wood products industry. AWC develops engineering data, technology, and standards on structural wood products for use by design professionals, building officials, and wood products manufacturers. AWC also provides technical, legal, and economic information on wood design, green building, and manufacturing environmental regulations.

American Zinc Association (AZA): AZA supports and advances zinc products and markets through research, development, technology transfer, and communication of the unique attributes that make zinc sustainable and essential for life. The association's main program areas are technology and market development; environment, health, and sustainability; and communications. AZA is affiliated with the International Zinc Association (IZA).

APA–The Engineered Wood Association: Founded in 1933 as the Douglas Fir Plywood Association, and later recognized as the American Plywood Association, APA changed its name to APA–The Engineered Wood in 1994. The association represents engineered wood manufacturers. It provides services such as quality testing, product research, market development for its members.

Appalachian Hardwood Manufacturers, Inc. (AMHI): AHMI promotes the use of logs, lumber, and products from the Appalachian Mountain region. It will assist users to make decisions relating to hardwoods.

Architectural Precast Association (APA): Members include manufacturers and their suppliers of architectural precast concrete products. The association establishes and upholds quality assurance for member products.

Architectural Woodwork Institute (AWI): A trade association that represents architectural woodworkers, suppliers, design professionals, and students from around the world.

Architectural Woodwork Manufacturers Association of Canada (AWMAC): In many ways the Canadian version of AWI.

Art Glass Association (AGA): An international organization whose purpose is to create awareness, knowledge, and involvement in the art glass industry.

ASIS International: ASIS International was originally the American Society for Industrial Security (ASIS); the organization changed its name in 2002. ASIS develops educational programs and materials that address broad security interests.

Asphalt Emulsion Manufacturers Association (AEMA): An international trade association that represents the asphalt emulsion industry. AEMA's goal is to expand the use and applications of asphalt emulsions.

Asphalt Institute (AI): An international trade association of petroleum asphalt producers, manufacturers, and affiliated businesses. AI promotes the use of petroleum asphalt, through engineering, research, marketing, and educational activities, and through the resolution of issues affecting the industry.

Asphalt Pavers Alliance (APA): A coalition of the Asphalt Institute, the National Asphalt Pavement Association, and the State Asphalt Pavement Associations. Its mission is to establish asphalt pavement as the preferred pavement through research, technology transfer, engineering, education, and innovation.

Asphalt Roofing Manufacturers Association (ARMA): A trade association that represents North American asphalt roofing manufacturers and their raw material suppliers.

Asphalt Recycling and Reclaiming Association (ARRA): The association promotes the recycling of existing roadway materials through various construction methodologies.

Associated Air Balance Council (AABC): The council establishes industry standards for the field measurement and documentation of HVAC systems. It also provides education, technical training, and certification for its members.

Associated General Contractors of America (AGC): The association serves construction professionals through the education of its members and students (scholarship programs), research, recognition programs, and charities for needy.

Associated Landscape Contractors of America (ALCA): A trade association that promotes business management skills and the profitability of its members' businesses.

Associated Locksmiths of America, Security Professionals Association, Inc. (ALOA): The association has recently changed its name from simply ALOA because it broadened its focus beyond "traditional" locksmithing. The association also encompasses electronic locksmiths, automotive locksmiths, and other security professions related to access control.

Associated Specialty Contractors (ASC): An umbrella organization of nine national associations of construction specialty contractors.

(International) Association of Foundation Drilling (ADSC): A trade association that advances the interests of people who are engaged in the design, construction, equipment manufacture, and distribution of anchored earth retention, drilled shaft, micro-piling, and related industries.

Association of Iron and Steel Technology (AIST): The association was formed in January 2004 from a merger of the Iron and Steel Society and the *Association of Iron and Steel Engineers.* It advances the technical development, production, processing, and application of iron and steel.

Association of Millwork Distributers (AMD): The association provides leadership, certification, education, promotion, networking, and advocacy to, and for, the millwork distribution industry.

Association of the Nonwoven Fabrics Industry (INDA): INDA focuses on networking events to help members increase sales and market share. It is a source for education, market leading data, global forecasts, testing standards, and trend reports.

Association of Physical Plant Administrators (APPA): Organized originally as the Association of Superintendents of Buildings and Grounds, the association later became the Association of Physical Plant Administrators of Universities and Colleges. In 1991, the name APPA: The Association of Higher Education Facilities Officers was adopted. In 2005, the association began to identify itself simply as APPA. It is a professional organization of college and university physical plant administrators, architects, and engineers.

Association of Pool and Spa Professionals (APSP): APSP is a trade organization that is dedicated to the growth and development of its members' businesses. It promotes the enjoyment and safety of pools and spas.

Association of Professional Landscape Designers (APLD): An international organization that advances the profession of landscape design and promotes the recognition of landscape designers as qualified and dedicated professionals.

Association of Residential Cleaning Services International (ARCSI): A trade association that comprises both large and small residential cleaning companies.

Association of State Floodplain Managers (ASFPM): An organization of professionals who are involved in floodplain management, flood hazard mitigation, the National Flood Insurance Program, and flood preparedness, warning, and recovery.

Association of the Wall and Ceiling Industry (AWCI): A trade organization that provides services and undertakes activities that enhance its members' ability to operate a successful business.

Association of Zoos and Aquariums (AZA): AZA provides the standards and best practices needed for animal care, wildlife conservation and science, conservation education, the guest experience, and community engagement.

ASTM International: Formerly, the American Society for Testing and Materials (ASTM), ASTM International is an organization that establishes material standards.

Audio Engineering Society (AES): The society is devoted exclusively to audio technology. It is an international organization that unites audio engineers, creative artists, scientists, and students worldwide by promoting advances in audio and disseminating new knowledge and research.

Automatic Fire Alarm Association (AFAA): A trade organization that supports business advancement, code development, and development of training and educational programs.

Barre Granite Association, Inc.: It has been estimated that one-third of the public and private monuments and mausoleums in America are products of the Barre quarries and Barre's "international" community of sculptors, artisans, mechanics, and laborers.

Bath Enclosures Manufacturers Association (BEMA): The association represents industry manufacturers, suppliers, and dealers in the United States and Canada.

BC Wood: A trade association that supports British Colombia's wood products manufacturers.

Blow in Blanket Contractors Association (BIBCA): An industry support association. BIBCA requires members to abide by a strict code of ethics.

Brick Industry Association (BIA): BIA represents distributors and manufacturers of clay brick and suppliers of related products and services to regulators and legislators. BIA comprises regions that manage programs in the Midwest/Northeast, Southeast, and Southwest.

Builders Hardware Manufacturers Association (BHMA): A trade association for North American manufacturers of commercial builder's hardware. BHMA is involved in standards, code, and life safety regulations and other activities that specifically impact builders' hardware.

Building Environment and Thermal Envelope Council (BETEC): Part of the National Institute of Building Sciences, an organization representing government and industry, BETEC is involved in communicating government policy and influencing standards development within the industry.

Building Officials and Code Administrators International, Inc. (BOCA): One of the three model code groups in the United States that has now merged into the International Code Council. This code is a minimum model regulatory code for the protection of public health, safety, welfare, and property by regulating and controlling the design, construction, quality of materials, use, occupancy, location, and maintenance of all buildings and structures within a jurisdiction. The code is used primarily in the North Central and Northeast United States.

Building Owners and Managers Association International (BOMA): BOMA represents the owners and managers of all commercial property types. It advances the interests of the entire commercial real estate industry through advocacy, education, research, standards, and information.

Building Research Establishment (BRE): BRE's home is in Watford, UK. BRE is an independent and impartial, research-based consultancy, testing, and training organization that offers expertise in all aspects of the built environment and associated industries.

Building Service Contractors Association International (BSCAI): A trade association of the building service industry, BSCAI members provide cleaning, facility maintenance, and other related services. The association provides contractor-specific educational programs, individual certifications, publications, a members-only purchasing program, seminars, industry data, and research and networking opportunities.

Building Seismic Safety Council (BSSC): An independent, voluntary organization that represents a wide range of building-related interests that are concerned with seismic safety. BSSC is a council of the National Institute of Building Sciences.

Building Stone Institute (BSI): The institute promotes and advances the use of natural stone. BSI provides its members with knowledge, information, products, and services for the design community and end user.

Business and Institutional Furniture Manufacturer's Association (BIFMA): A trade organization, it advocates, informs, and develops standards for the North American office and institutional furniture industry.

Cable Tray Institute (CTI): The institute supports the cable tray industry by engaging in research, development, education, and the dissemination of information designed to promote, enhance, and increase the visibility of the industry.

California Manufactured Housing Institute (CHMI): A professional and trade association that represents builders of factory-constructed homes, retailers, financial services, developers, and community owners and their supplier companies.

California Redwood Association (CRA): An industry advocacy association that provides information on the use of redwood and its sustainability, "how-to" advice, etc.

Canadian Carpet Institute (CCI): An organization that represents Canada's carpet manufacturers and their suppliers.

Canadian Copper and Brass Development Association (CCBDA): A communications and advisory group for the copper industry. CCBDA represents and gives support to its members and users of copper and copper alloys, including educators and the general public.

Canadian Disaster Restoration Group (CDRG + REDTEAM): Established in 2004, CDRG + REDTEAM is a team of restorers, property specialists, and project managers who provide property restoration services throughout Canada.

Canadian Hardwood Plywood Veneer Association (CHPVA): A national association that represents the Canadian hardwood plywood and veneer industry with technical, regulatory, quality assurance, and product acceptance.

Canadian Institute of Steel Construction (CISC): The institute promotes good design, safety, and the efficient, economical, and sustainable use of structural steel.

Canadian Plywood Association (CANPLY): The association carries a complete line of products from the leading manufacturers. Certiwood™, a part of

CANPLY, is an engineered wood products testing and certification agency.

Canadian Roofing Contractor's Association (CRCA): CRCA consists of companies that are actively engaged in Canada in the roofing and related sheet metal contracting business, along with companies engaged in manufacturing or supplying materials and services that are used in any branch of the roofing and sheet metal industry.

Canadian Security Association (CANASA): An organization dedicated to the advancing the security industry and supporting security professionals in Canada.

Canadian Sheet Steel Building Institute (CSSBI): An industry association that is responsible for the development and dissemination of industry standards. CSSBI is a technical information and resources expert for both the general public and sheet steel manufacturers.

Canadian Society of Landscape Architects (CSLA): CSLA is dedicated to advancing the art, science, and business of landscape architecture.

Canadian Standards Association (CSA): CSA is the Canadian equivalent of ASTM.

Canadian Steel Producers Association (CSPA): The association is committed to a strong and internationally competitive Canadian steel sector.

Canadian Wood Council (CWC): A trade organization that represents manufacturers of Canadian wood products that are used in construction.

Carpet Cushion Council (CCC): CCC educates carpet retailers, manufacturers, distributors, and cushion manufacturers about the benefits of carpet cushion.

Carpet and Rug Institute (CRI): The institute provides science-based facts about carpeting and rugs.

Cast Iron Soil Pipe Institute (CISPI): A trade organization that provides technical reports to advance interest in the manufacture, use, and distribution of cast iron soil pipe and fittings. CISPI strives to improve the industry's products, achieve standardization, and provide a continuous program of product testing, evaluation, and development.

Cast Stone Institute (CSI): A self-governing association of producers and suppliers to the cast stone industry. CSI is a spokesperson for cast stone and provides counsel to the architectural and engineering communities.

Cedar Shake and Shingle Bureau (CSSB): An organization that promotes the use of Certi-label™ cedar roofing and sidewall products.

Ceiling and Interior Systems Construction Association (CISCA): The association serves the acoustical and specialty ceilings and interior finishes industry. CISCA provides networking, education, resources, and technical guidelines to its members.

Cellulose Insulation Manufacturers Association (CIMA): A trade association for the cellulose segment of the thermal/acoustical insulation industry.

Cement Association of Canada (CAC): The association represents the Canadian cement firms that have clinker and cement manufacturing facilities, granulators, grinding facilities, and cement terminals.

Ceramic Glazed Masonry Institute (CGMI): A trade organization of glazed ceramic product manufacturers and distributors.

Ceramic Tile Distributors Association (CTDA): An international association of distributors, manufacturers, and allied professionals of ceramic tile and related products. CTDA connects, educates, and strengthens tile and stone distributors.

Ceramic Tile Institute of America (CTIOA): CTIOA provides manufacturer's information but does not test products or validate manufacturer's claims. The institute has recently collaborated with the Los Angeles County Metropolitan Transportation Authority (Metro) and noted artists to implement tile artworks.

Certified Floorcovering Installers (CFI): An organization that identifies, trains, and certifies flooring installers according to skill and knowledge, it also provides the industry with educational programs.

Chain Link Fence Manufacturers Institute (CLFMI): An organization comprising firms that manufacture chain link fence fabric, fittings, framework, accessories, and/or the materials used to produce them. CLFMI provides advice to people such as architects, engineers, and contractors regarding chain link fencing.

Chicago Roofing Contractors Association (CRCA): A local trade association of roofing and waterproofing contractors in the greater Chicago area. Contractors, manufacturers, distributors, manufacturers, representatives, and consultants are members.

Chimney Safety Institute of America (CSIA): The institute fosters public awareness of issues relating to chimney and venting performance and safety. CSIA promotes technical training and certification opportunities.

Cleaning and Restoration Association (CRA): A trade organization of cleaning and restoration firms from the Western United States. CRA recognizes the IICRC (Institute of Inspection, Cleaning, and Restoration Certification) certification programs and standards

Cold-Formed Steel Engineers Institute (CFSEI): The institute comprises structural engineers and other design professionals who provide designs for commercial and residential structures with cold-formed steel.

Commercial Food Equipment Service Association (CFESA): A trade association of professional service and parts distributors.

Composite Panel Association (CPA): CPA represents the North American composite panel and decorative surfacing industries. CPA sponsors all ANSI standards related to particleboard, MDF, hardboard, and engineered wood siding and trim, as well as CPA's Eco-Certified Composite™ sustainability standard and certification program.

Compressed Air and Gas Institute (CAGI): The institute serves the compressed air industry by providing technical, educational, and promotional support and being involved in other matters that affect the industry.

Compressed Gas Association (CGA): CGA promotes the safe, secure, and environmentally responsible manufacture, transportation, storage, trans-filling, and disposal of industrial and medical gases and their containers.

Concrete Anchor Manufacturers' Association (CAMA): Members include manufacturers of concrete anchoring systems and associated technical representatives. CAMA works with code organizations to develop uniform codes and standards and advance the use of anchoring systems.

Concrete Countertop Institute (CCI): A trade organization that provides the concrete countertop industry with training, membership programs, and advocacy, consultation, and guidance.

Concrete Foundation Association (CFA): A trade association that provides promotional materials, educational seminars, networking opportunities, and technical and informative meetings for contractors who are in the residential concrete foundation industry.

Concrete Polishing Association of America (CPPA): The association builds and maintains standards and is an advocate for concrete that is processed to a polished finish.

Concrete Reinforcing Steel Institute (CRSI): A national trade association. It is a resource for information related to steel-reinforced concrete construction. CRSI industry members include manufacturers, fabricators, and placers of reinforcing bars and related products.

Concrete Sawing and Drilling Association (CSDA): The association promotes the use of professional specialty sawing and drilling contractors and their methods.

Confederation of International Contractors Association (CICA): A trade organization, CICA is dedicated to improving the business environment in the construction industry. The association promotes investment in engineering and building that enhances both our environment and the quality of life for all.

Construction Engineering Research Laboratory (CERL): Part of the U.S. Army Engineer Research and Development Center (USAERDC) that is the integrated Army Corps of Engineers' research and development organization, CERL conducts research and development in infrastructure and environmental sustainment.

Construction Management Association of America (CMAA): A North American organization that is dedicated exclusively to the interests of professional construction and program management.

Construction Materials Recycling Association (CMRA): CMRA promotes the safe and economically feasible recycling of recoverable construction and demolition (C&D). These materials include aggregates such as concrete, asphalt, asphalt shingles, gypsum wallboard, wood, and metals.

Construction Specifications Institute (CSI): The institute advances building information management and education of project teams to improve facility performance.

The CSI's Master Format is a system of numbers and titles for organizing construction information into a regular, standard order or sequence. By establishing a master list of titles and numbers, Master Format promotes standardization and thereby facilitates the retrieval of information and improves construction communication. It provides a uniform system for organizing information in project manuals, for organizing project cost data, and for filing product information and other technical data.

Consumer Credit Counseling Service (CCCS): A nationwide, nonprofit organization that helps consumers get out of debt and improve their credit profile.

Consumer Electronics Association (CEA): The association unites companies within the consumer technology industry. Members receive market research, networking, educational programs and technical training, and advocacy.

Continental Automated Buildings Association (CABA): An international industry association that is dedicated to the advancement of intelligent home and intelligent building technologies. Membership companies are involved in the design, manufacture, installation, and retailing of products relating to home automation and building automation. Public organizations, including utilities and government, are also members.

Conveyor Equipment Manufacturers Association (CEMA): A trade association that serves North American manufacturers and designers of conveyor equipment. CEMA is focused on voluntary adherence to design standards, safety, manufacture, and applications to promote industry growth.

Conveyor Section of the Material Handling Institute: The Materials Handling Institute (MHI) is a material handling, logistics, and supply chain association that offers education, networking and solution sourcing for members; the Conveyor Section is part of the MHI. See Materials Properties Council (MPC).

Cold Regions Research and Engineering Laboratory (CRREL): A U.S. Army Corps of Engineers Laboratory, it advances and applies science and engineering to complex environments, materials, and processes in all seasons and climates, with unique core competencies related to the Earth's cold regions.

Cool Metal Roofing Coalition: The coalition has as its mission to educate architects, building owners, designers, code and standards officials, and other stakeholders about the sustainable, energy-related benefits of cool metal roofing.

Cool Roof Rating Council (CRRC): An organization that maintains a third-party rating system for the radiating properties of roof surfacing materials. See Cool Metal Roofing Coalition.

Cooling Technology Institute (CTI): CTI advocates and promotes the use of environmentally responsible evaporative heat transfer systems (EHTS), cooling towers, and cooling technology through education, research, standards development and verification, government relations, and technical information exchange.

Copper Development Association (CDA): CDA supports the copper industry with market development, engineering, and information services.

Council for Interior Design Accreditation (CIDA): The council ensures a high level of quality in interior design education through three primary activities: setting standards for postsecondary education, evaluating and accrediting colleges and universities, and facilitating outreach and collaboration.

Council of American Building Officials (CABO): CABO has joined with the **Southern Building Code Congress International (SBCCI)** and the **International Code Council (ICC)**, and they now write the internationally recognized One and Two Family Dwelling Code. These codes provide administrative and technical directions for all phases of residential construction.

Council of Forest Industries (COFI): The council works with governments, communities, organizations, and individuals to ensure that forest policies in British Columbia support the forest sector.

Council of Landscape Architectural Registration Boards (CLARB): The council is dedicated to ensuring that all people who practice landscape architecture are fully qualified. Its members include the licensure boards in 48 states, two Canadian provinces, and the territory of Puerto Rico.

Custom Electronic Design and Installation Association (CEDIA): An international trade association of companies that specialize in planning and installing electronic systems for the home.

Dade County: A Florida County, including Miami, that has set numerous standards and requirements for hurricane-resistant windows and doors. I don't think so. Use your judgement.

Decorative Plumbing and Hardware Association (DPHA): An organization that represents independent retailers, manufacturers, and manufacturer's representatives. DPHA develops programs and publications to improve business practices, employee performance, and quality of service.

Deep Foundation Institute (DFI): The institute helps its members to improve in all aspects of planning, designing, and constructing deep foundations and deep excavations.

Door and Access Systems Manufacturers Association International (DASMA): A North American association of manufacturers of garage doors, rolling doors, high performance doors, garage door operators, vehicular gate operators, and access control products.

Door and Hardware Institute (DHI): DHI is dedicated to the architectural openings industry. It advances life safety and security within the built environment, and is an advocate and resource for information, professional development, and certification.

Dry Stone Conservancy (DSC): The conservancy is dedicated to preserving dry-laid stone structures and promoting the ancient craft of dry stone masonry.

Ductile Iron Pipe Research Association (DIPRA): An association that is supported by ductile iron pressure pipe manufacturers in North America.

Earthquake Engineering Research Institute (EERI): A technical society of engineers, geoscientists, architects, planners, public officials, and social scientists. EERI members include researchers, practicing professionals, educators, government officials, and building code regulators.

Edison Electric Institute (EEI): An association that represents all U.S. investor–owned electric companies. Its members provide electricity to 220 million Americans in 49 states and the District of Columbia. EEI has 70 international electric companies as affiliate members and 250 industry suppliers and related organizations as associate members.

Electrical Generating Systems Association (EGSA): The association is exclusively dedicated to on-site power generation. The association comprises manufacturers, distributor/dealers, contractors/integrators, manufacturer's representatives, consulting and specifying engineers, service firms, end-users, and others who make, sell, distribute, and use on-site power generation technology and equipment, including generators, engines, switchgear, controls, voltage regulators, governors, and much more.

Electronic Security Association (ESA): A professional trade association that represents the electronic life safety, security, and integrated systems industry.

Elevator Escalator Safety Foundation (EESF): An organization that was created by the elevator/escalator industry to develop and disseminate safety materials to the public in order to eliminate preventable accidents on the industry's equipment.

Engineered Wood Association (EWA): The association tests and sets standards for all varieties of plywood used in the United States.

EnOcean Alliance: The EnOcean Alliance develops and promotes self-powered wireless monitoring and control systems for sustainable buildings by formalizing the interoperable wireless standard.

Environmental Information Association (EIA): EIA began as the National Asbestos Council. EIA is a trade organization that focuses on the environmental health and safety industry. The association collects, generates, and disseminates information pertaining to environmental health hazards to occupants of buildings, industrial sites, and other facility operations.

(The United States) Environmental Protection Agency (EPA): An agency of the U.S. federal government that was created for the purpose of protecting human health and the environment by writing and enforcing regulations based on laws passed by Congress.

Erosion Control Technology Council (ECTC): The council consists of a broad range of professions and specialties, including site engineers, consultants, regulatory agencies, earthwork and seeding contractors, erosion control product suppliers, and manufacturers. ECTC has set as its mission to be the recognized industry authority in the development of standards, testing, and installation techniques for rolled erosion control products (RECPs), hydraulic erosion control products (HECPs) and sediment retention fiber rolls (SRFRs).

Ethylene Propylene Diene Monomer (EDPM) Roofing Association (ERA): A trade association in support of the EDPM industry.

Expanded Metal Manufacturers Association (EMMA): A division of the **National Association of Architectural Metal Manufacturers (NAAMM).**

Expanded Shale, Clay and Slate Institute (ESCSI): An association for manufacturers of rotary kiln-produced expanded shale, expanded clay, and expanded slate lightweight aggregate.

Exterior Design Institute (EDI): EDI was founded to train and certify building envelope and EIFS (exterior insulation and finish systems) inspectors and moisture analysts. It promotes quality control within the construction industry.

Exterior Insulation and Finishing System (EIFS) Industry Members Association: A trade association that supports the industry by developing consensus technical, training, installation, and design standards for use by architects, designers, code bodies and officials, and other technical associations, by monitoring and positively influencing government actions, by working to assure the long-term availability of qualified contractors, and by providing other member services as appropriate.

EuroWindoor: A consortium of European window, door and curtain wall industry associations that are involved in the development of common EU standards.

Federal Emergency Management Agency (FEMA): FEMA supports U.S. citizens and first responders to disasters. It also builds, sustains, and improves our nation's capability to prepare for, protect against, respond to, recover from, and mitigate all hazards.

Federal Housing Authority (FHA): The FHA sets construction standards throughout the United States.

Fenestration Canada: A trade organization that was formerly called the **Canadian Window and Door Association (CWDMA).** Fenestration Canada represents and supports all aspects of the window and door manufacturing industry, including formulating and promoting standards of quality in manufacturing, design, marketing, distribution, sales, and application of all types of window and door products.

Fiberglass Tank and Pipe Institute: The Fiberglass Tank and Pipe Institute provides a forum through which the fiberglass-reinforced thermoset plastic (RTP) industry can advance the use of fiberglass products that are used in the underground tank and piping marketplace. Fiberglass Tank and Pipe Institute coordinates market studies, gathers statistics, and provides standard-setting organizations with technical data and it disseminates information to the government, industry, and the public.

Finishing Contractors Association (FCA): A trade organization that provides programs, products, and services and establishes relationships with other relevant organizations.

Fire Equipment Manufacturers' Association (FEMA): Another FEMA. This FEMA is also committed to saving lives and protecting property. FEMA provides educational opportunities, advances best industry standards, and provides advocacy for the industry.

Fire Suppression Systems Association (FSSA): An international trade association comprising manufacturers, suppliers, and designer-installers, who are dedicated to providing a higher level of fire protection. **FSSA** members are specialists in protecting high value special hazard areas from fire.

Firestop Contractors International Association (FCIA): A trade organization of firestop contractors.

FLO-CERT GmbH: An independent International Certification company. FLO assists in the socio-economic development of producers in the Global South. FLO-CERT allows people to identify products that meet agreed upon environmental, labor, and development standards.

Floor Covering Installation Contractors Association (FCICA): The association provides a network for problem solving, education and support, to enhance its members' businesses, and the flooring industry.

Floor Installation Association of North America (FIANA): An organization whose members are from Canada and the United States. Members must be manufacturers or distributors of floor installation products and/or flooring accessories.

Foodservice Consultants Society International (FCSI): Members are consultants with competencies that span the entire food service industry.

Forest Products Laboratory (FPL): A laboratory of the U.S. Department of Agriculture, Forest Service, Research and Development.

Forest Products Society: The Forest Products Society is an international association that provides information network for all segments of the forest products industry. Members represent both private and public research and development, industrial management and production, marketing, education, government, engineering, and consulting. The society also functions as the distributor for the technical publications of the American Wood Council.

Forest Stewardship Council (FSC): An independent organization that protects forests for future generations. FSC sets standards under which forests and companies are certified to demonstrate that they are managing their forests responsibly. FSC's membership consists of three equally weighted chambers: environmental, economic, and social, to ensure these are balanced and the highest level of integrity. Members are in regular contact with their peers, customers, and suppliers.

Gas Technology Institute (GTI): A research, development, and training organization that addresses energy and environmental issues.

GeoExchange® (GEO): A trade association that promotes the manufacture, design, and installation of GeoExchange® systems. GEO supports its members' business objectives while promoting sustainable growth of the geothermal heat pump industry.

Geosynthetic Institute (GSI): A consortium of organizations that are interested in, and involved with, geosynthetics: geotextiles, geomembranes, geogrids, geonets, geocomposites, geosynthetic clay liners, geopipe, geocells, and geofoam.

Geothermal Energy Association (GEA): A trade association that supports the expanded use of geothermal energy and the development of geothermal resources for electrical power generation and direct-heat uses.

Geothermal Resources Council (GRC): An educational association that serves as a focal point for continuing professional development for its members through its outreach, information transfer, and education services.

Glass Association of North America (GANA): GANA places members in regular contact with their peers, customers, and suppliers. The association also provides members with educational programs, publications, networking opportunities, meetings, and conventions.

Green Roofs for Healthy Cities – North America Inc. (GRHC): GRHC promotes the industry throughout North America.

Gypsum Association (GA): A trade association that promotes the use of gypsum in the United States and Canada on behalf of its member companies.

Hardwood Manufacturers Association (HMA): A national trade organization with membership limited to hardwood sawmills and lumber concentration yards located in the United States.

Hardwood Plywood and Veneer Association (HPVA): A trade association that represents the interests of the hardwood plywood, hardwood veneer, and engineered hardwood flooring industries.

Hearth, Patio, and Barbecue Association (HPBA): In 2002, the Hearth Products Association (HPA) merged with the Barbecue Industry Association (BIA) to form HPBA. It is an international trade association that includes manufacturers, retailers, distributors, manufacturers' representatives, service and installation firms, and other companies and individuals.

Heat Exchange Institute (HEI): A trade association that is committed to the technical advancement, promotion, and understanding of a broad range of utility and industrial-scale heat exchange and vacuum apparatus.

Heating, Air-conditioning and Refrigeration Distributers International (HARDI): The association's members market, distribute, and support heating, air-conditioning, and refrigeration equipment, parts and supplies. HARDI distributor members serve installation and service/ replacement contractors in the residential, commercial, industrial, and institutional markets.

Heating, Refrigeration, and Air Conditioning Institute of Canada (HRAI): A national association that represents heating, ventilation, air conditioning and refrigeration (HVACR) manufacturers, wholesalers, and contractors and provides information about HVACR to Canadians.

Hollow Metal Manufacturers Association (HMMA): The association promotes the advantages of hollow metal products. It is a division of the National Association of Architectural Metal Manufacturers (NAAMM).

Home Fire Sprinkler Coalition (HFSC): HFSC is a charitable organization. It provides independent, noncommercial information about residential fire sprinklers. HFSC offers educational material with details about installed home fire sprinkler systems, how they work, why they provide affordable protection, and answers to common myths and misconceptions about their operation.

Home Furnishings Independents Association (HFIA): A trade organization for member businesses.

Home Ventilating Institute (HVI): The institute certifies a wide range of home ventilating products that are manufactured by companies located throughout the world.

Hydraulics Institute (HI): An association of pump industry manufacturers, HI provides product standards and a forum for the exchange of industry information.

Illuminating Engineering Society (IES): The society is dedicated to promoting the art and science of quality lighting to its members, allied professional organizations, and the public.

INDA: See Association of the Nonwoven Fabrics Industry.

Independent Electrical Contractors (IEC): A national trade association for merit shop electrical and systems contractors. IEC develops and fosters a high level of quality and services within the industry.

Independent Office Products and Furniture Dealers Association (IOPFDA): A trade association for North American independent dealers of office products and office furniture. IOPFDA concentrates on providing independent dealers with information, tools, and knowledge needed to run their businesses.

Indiana Limestone Institute (ILI): A resource for architects, contractors, building owners, and others seeking information about the use of Indiana Limestone in construction.

Indoor Air Quality Association (IAQA): The association promotes uniform standards, procedures, and protocols in the Indoor Air Quality industry. IAQA's membership is consolidated with The American Council for Accredited Certifications (ACAC) and the Indoor Environmental Standards Organization (IESO).

Industrial Fabrics Association International (IFAI): A trade association comprising member companies representing global specialty fabrics.

Industrial Fasteners Institute (IFI): A globally recognized, North American-focused, association that represents manufacturers of mechanical fasteners and formed parts and suppliers to the industry.

Industrial Perforators Association (IPA): A highly specialized production resource for punching very large numbers of holes in a wide variety of materials. Hole sizes range from a few thousandths of an inch in diameter up to more than 3 inches, while the materials that can be perforated can be as thin as foil or as thick as a 1-inch steel plate.

InfoComm International (InfoComm): A trade association that represents the professional audiovisual and information communications industries worldwide.

Innovative Pavement Research Foundation (IPRF): The foundation develops strategies and implements programs of research, technology advancement and transfer, and public education regarding concrete highways, streets, roads, and airports. IPRF is sponsored by the American Concrete Pavement Association.

Institute of Electrical and Electronic Engineers (IEEE): IEEE fosters technological innovation and excellence for the benefit of humanity. IEEE is pronounced "Eye-triple-E."

Institute of Fire Engineers (IFE): The institute promotes, encourages, and improves all aspects of the science and practice of fire engineering, fire prevention, and fire extinction.

Institute of Heating and Air Conditioning Industries, Inc. (IHACI): A trade association of contractors, manufacturers, distributors, utility firms, and related businesses actively engaged in the heating, ventilation, air conditioning, refrigeration, and sheet metal industries.

Institute of Inspection Cleaning and Restoration Certification (IICRC): IICRC identifies and promotes an international standard of care that establishes and maintains the health, safety and welfare of the built environment. **It is a certification and standard-setting organization for the inspection, cleaning, and restoration industries.**

Institute of Noise Control Engineering (INCE/USA): A professional organization whose primary purpose is to promote engineering solutions to environmental, product, machinery, industrial, and other noise problems. INCE/USA is a member society of the International Institute of Noise Control Engineering.

Institute of Transport Engineers (ITE): ITE is an international educational and scientific association of transportation professionals who are responsible for meeting mobility and safety needs. ITE facilitates research, planning, functional design, implementation, operation, policy development, and management for any mode of ground transportation.

Institute of the Ironworking Industry (III): A labor-management trade association that protects, promotes, fosters, and advances the unionized erection industry.

Insulated Cable Engineers Association (ICEA): An association whose members are sponsored by many of North America's cable manufacturers, ICEA is dedicated to developing cable standards for the electric power, control, and telecommunications industries.

Insulating Glass Certification Council (IGCC): A trade organization for insulating glass unit manufacturers, consumers, specifiers, and others who are stakeholders in this industry.

Insulating Glass Manufacturers Alliance (IGMA): A trade organization comprising certified insulating glass manufacturers, their suppliers and associates, window manufacturers, representatives from the architectural community, energy efficiency lobbies, code officials, and others interested in the design and long-term performance of insulating glass units.

Insulation Contractors Association of America (ICAA): A trade organization that represents professional residential and commercial contractors.

Interior Design Educators Council® (IDEC): The council advances interior design education, scholarship, and service.

Interlocking Concrete Pavement Institute (ICPI): A trade association that represents the industry. Membership consists of interlocking paver manufacturers, design professionals, paver installation contractors, and suppliers of products and services related to the industry.

(United States) Internal Revenue Service (IRS): The organization that publishes all of the United States' tax laws and rules and forms that are associated with them.

International Association of Amusement Parks and Attractions (IAAPA): An international trade association for permanently situated amusement facilities worldwide.

International Association of Certified Home Inspectors (InterNACHI): An international trade organization that comprises mainly independent inspectors. InterNachi provides education, training, certification, benefits, and support for its members; inspectors are required to complete 24 hours of continuing education annually.

International Association of Electrical Inspectors (IAEI): The association promotes safe products and safe installations. Members include electrical inspectors, testing agencies, standards organizations, manufacturers, distributors, installers, and contractors.

International Association of Foundation Drilling (ADSC-IAFD): A trade association comprising people, firms, and corporations engaged in the design, construction, equipment manufacture, and distribution in and for the anchored earth retention, drilled shaft, micropiling, and corresponding industries.

International Association of Lighting Designers (IALD): IALD promotes the advancement and recognition of independent, professional lighting designers.

InterNational Association of Lighting Management Companies (NALMCO®): The association establishes and promotes professional standards for lighting management professionals through education, representation, the enhancement of professionalism and distribution of information about the industry.

International Association of Plumbing and Mechanical Officials (IAPMO): IAPMO provides code development assistance, education, plumbing and mechanical product testing and certification, and building product evaluation and a quality assurance program. The association publishes standards for mechanical products covering heating, ventilation, cooling, and refrigeration system products; members contribute to the development of the Uniform Mechanical Code. The association also publishes standards covering products used in the recreational vehicle and manufactured housing industry called IAPMO Trailer Standards.

International Association of Professional Security Consultants (IAPSC): The association establishes and maintains standards for professionalism and ethical conduct in the industry.

The International Cast Polymer Alliance of the American Composites Manufacturers Association (ICPA): The alliance represents manufacturers, suppliers, fabricators, and installers of cast polymer composites, including cultured marble, cultured granite, cultured onyx, and solid surface kitchen and bath products.

International Code Council (ICC): ICC publishes: the **International Building Code** that has been adopted throughout most of the United States; the **International Energy Conservation Code** (IECC): that sets forth compliance methods for energy-efficient construction of both residential and nonresidential construction; and the **International Residential Code** (IRC) that primarily covers low-rise residential construction.

International Concrete Repair Institute (ICRI): The institute serves to improve the quality of concrete restoration, repair and protection, through education of, and communication among, the members and those who use their services.

International Cost Engineering Council (ICEC): ICEC promotes cooperation between national and multinational cost engineering, quantity surveying, and project management organizations worldwide for their mutual well-being and that of their individual members.

International Council of Building Officials (ICBO): One of the three model code groups in the United States that has merged to form the International Code Council.

International Dark Sky Association (IDA): IDA is a recognized authority on light pollution. The association promotes "light what you need, when you need it." IDA works with manufacturers, planners, legislators, and citizens to provide energy-efficient options that direct the light where you want it to go, not up into the sky.

International Door Association (IDA): The association provides programs and services to door and access systems dealers whose service products include residential and commercial doors and operators and fire doors and gates.

InterNational Electrical Testing Association (NETA): The association serves the electrical testing industry by establishing standards, publishing specifications, accrediting independent, third-party testing companies, certifying test technicians, and promoting the professional services of its members. NETA also collects and disseminates information and data to the electrical industry and educates the public and end user about electrical acceptance and maintenance testing.

International Erosion Control Association (IECA): IECA is devoted to helping members solve the problems caused by erosion and its byproduct sediment.

International Facility Management Association (IFMA): It is a widely recognized international association for facility management professionals. IFMA certifies facility managers, conducts research, and provides a wide range of educational courses and is a leading voice in the industry.

International Firestop Council (IFC): A trade association of manufacturers, distributors, and installers of passive fire protection materials and systems in North America. IFC promotes the technology of fire and smoke containment in modern building construction through research, education, and development of safety standards and code provisions.The organization used firestop.

International Furnishings and Design Association (IFDA): IFDA brings together professionals in the furnishing and design industries through networking, education, and professional development.

International Ground Source Heat Pump Association (IGSHPA): A not-for-profit organization that advances ground source heat pump (GSHP) technology on local, state, national, and international levels.

International Hurricane Protection Association (IHPA): IHPA is involved in all issues that affect the hurricane protection industry. IHPA brings together suppliers, manufacturers, contractors, engineers, architects, code writers, and government officials.

International Institute of Noise Control Engineering (I-INCE): I-INCE is a worldwide consortium of organizations concerned with noise control, acoustics, and vibration. The primary focus of the institute is on unwanted sounds and on vibrations producing such sounds when transduced.

International Institute of Welding (IIW): A international scientific and engineering institute that focuses on welding, brazing, and related technologies. Its membership consists of the national welding societies from around the world.

International Interior Design Association (IIDA): The association provides a forum to demonstrate design professionals' impact on the health, safety, well-being, and virtual soul of the public. IIDA strives to balance good design and best business practices.

International Masonry Institute (IMI): IMI offers training for craftworkers, professional education for masonry contractors and free technical assistance to the design and construction communities. IMI is an alliance between the International Union of Bricklayers and Allied Craftworkers (BAC) and their signatory contractors.

International Organization for Standardization (ISO): The association certifies a company's ability to consistently manufacture quality products to ISO standards (ISO 9000, 9001, etc.).

International Parking Institute (IPI): The institute's members include professionals from cities, port authorities, civic centers, academic institutions, hospitals and healthcare facilities, airports, corporate complexes, race tracks, transit and transportation agencies, retail, hospitality, and entertainment and sports centers, as well as architects, engineers, financial consultants, urban planners, and suppliers of equipment, products and services to the parking and transportation industries.

International Play Equipment Manufacturers Association (IPEMA): The association provides third-party product certification services for U.S. and Canadian public play equipment and public play surfacing materials in the United States. IPEMA serves IPEMA-certified member companies, affiliated playground industry groups, and anyone with an interest in playground equipment regulations.

International Safety Equipment Association (ISEA): An association dedicated to personal protective equipment and technologies. Its members design, manufacture, test, and use protective clothing and equipment.

International Sign Association (ISA): The association provides its members with information about current engineering research, EPA compliance issues, and other relevant matters.

International Slurry Surfacing Association (ISSA): An international trade association comprising contractors, equipment manufacturers, public officials, research personnel, consulting engineers, and other industry professionals. ISSA promotes the concept of pavement preservation. ISSA provides members with information, technical assistance, and opportunities for networking and professional development.

International Society of Arboriculture (ISA): ISA promotes arboriculture and fosters an awareness of the benefits of trees through research, technology, and education.

International Staple, Nail & Tool Association (ISANTA): An international organization of premier power fastening companies that are involved in the designing, manufacturing, and selling of pneumatic and cordless tools and the fasteners they drive.

International Surface Fabricators Association (ISFA): ISFA certifies member contractors who fabricate and install countertops.

International Tropical Timber Organization (ITTO): The organization's international membership is committed to achieving exports of tropical timber and timber products from sustainably managed sources. ITTO assists governments, industry, and communities to manage their forests and add value to their forest products, and to maintain and increase the transparency of the trade and access to international markets.

International Window Cleaning Association (IWCA): A trade organization that represents window cleaning companies to international, national, state, and local regulatory agencies and promotes the welfare of the industry through advocacy, education, training, and community involvement.

International Window Film Association (IWFA): A trade organization, IWFA partners with manufacturers and other members to increase consumer awareness and demand for all types of professionally installed window film products.

International Wood Products Association (IWPA): The association advances international trade in wood products by providing education and leadership in business, environmental, and public affairs.

International Zinc Association (IZA): An organization that is based in Brussels, Belgium. IZA is dedicated exclusively to the interests of zinc and its users by promoting such end uses as corrosion protection for steel and crop nutrition.

Intertek Testing Services – Warnock Hersey (ITS): Intertek tests to ensure products meet quality, health, environmental, safety, and social accountability standards.

Irrigation Association (IA): A trade organization for irrigation equipment and system manufacturers, dealers, distributors, designers, consultants, contractors, and end users.

Kitchen Cabinet Manufacturers Association (KCMA): A voluntary trade association representing North American cabinet manufacturers and suppliers to the industry. KCMA promotes responsible environment practices in the industry.

League of Hard Flooring Professionals: A fairly new organization, the League of Hard Flooring Professionals was created to provide a source of recognized experts in the hard floor industry for federal, state, and municipal agencies and legislative and judicial bodies. The league also provides education.

Lighting Controls Association (LCA): An association that is administered by the **National Electrical Manufacturers Association (NEMA)**. LCA is dedicated to educating the professional building design, construction, and management communities about the operation of automatic switching and dimming controls.

Lightning Protection Institute (LPI): The institute designs and develops information resources on complete lightning protection systems for consumers and designers. LPI also markets education products to members for use in the construction industry.

Lighting Research Center (LRC): A university-based research center (Rensselaer Polytechnic Institute) devoted to lighting. LRC offers graduate education in lighting, including one- and two-year master's programs and a PhD program. LRC also provides training programs for government agencies, utilities, contractors, lighting designers, and other lighting professionals.

Lighting Safety Alliance (LSA): A not-for-profit corporation comprising lightning protection manufacturers, distributors, and installers. LSA evaluates and responds to legislative, administrative, and regulatory issues facing the industry. Additionally, LSA serves as an informational clearinghouse for its membership.

Manufactured Housing Institute (MHI): A national trade organization that represents all segments of the factory-built housing industry. MHI provides industry research, promotion, education, and government relations programs.

Maple Flooring Manufacturers Association, Inc. (MFMA): A source of technical information about hard maple flooring, MFMA publishes grade standards, guide specifications, floor care recommendations, and specifications for athletic flooring sealers and finishes.

Marble Institute of America (MIA): A source of information on standards of natural stone workmanship and practice and the suitable application of natural stone products. MIA promotes stone usage in the commercial and residential marketplaces. Membership includes natural stone producers, exporters/importers, distributors/wholesalers, fabricators, finishers, installers, and industry.

Mason Contractors Association of America (MCAA): A trade association that represents mason contractors. MCAA provides continuing education, promotes codes and standards, fosters a safe work environment, recruits future tradespeople, and markets the benefits of masonry materials.

Masonry Advisory Council (MAC): MAC is dedicated to providing the public with general and technical information about masonry design and detailing. MAC has an on-line technical library and industry directory.

Masonry Heater Association of North America (MHA): An association of builders, manufacturers, and retailers of masonry heaters, MHA promotes the industry, sponsors research and development, shapes regulations, standards and codes, and provides information/education to the public and its members.

Masonry Institute of America (MIA): A trade organization that is primarily supported by Southern California union signatory masonry contractors through a labor management contract between unions and contractors. MIA does not practice architecture or engineering or sell masonry building materials, but it is active in the development and distribution of seminars and publications on the use of masonry.

The Masonry Society (TMS): Members are design engineers, architects, builders, researchers, educators, building officials, material suppliers, manufacturers, and other interested people. TMS gathers and disseminates technical information through its committees, publications, codes and standards, slide sets, videotapes, computer software, newsletter, refereed journal, educational programs, professors' workshop, scholarships, certification programs, disaster investigation team, and conferences.

Masonry Veneer Manufacturers Association (MVMA): An incorporated trade association that represents the manufactured stone veneer industry's manufacturing companies and their suppliers. MVMA advances the growth of the manufactured masonry veneer industry through proactive technical, advocacy, and awareness efforts.

Master Painters Institute (MPI): The institute is dedicated to the establishment of quality standards and quality assurance in the painting and coating application industries.

Material Handling Industry of America (MHI): The institute was formed as the Material Handling Institute to advance the interests of material handling and logistics companies, systems and software manufacturers, consultants, systems integrators and simulators, and third-party logistics providers and publishers. It changed its name to Material Handling Industry of America in the late 1980s; it continues to use both MHI as an abbreviation.

Materials Properties Council (MPC): The council was established in 1966 by the American Society of Mechanical

Engineers, ASM International, ASTM, and the Engineering Foundation. Industry, technical organizations, codes and standards developers, and government agencies support it.

Mechanical Contractors Association of America, Inc. (MCAA): An association of mechanical, plumbing, and service contractors. MCAA provides educational programs, a catalog of resources to help members manage and grow their businesses, periodicals, and other business services.

Medical Gas Professional Healthcare Organization, Inc. (MGPHO): The organization is made up of people and companies that are dedicated to advancing the safe design, manufacture, installation, maintenance, and inspection/verification of medical gas and vacuum delivery systems through education.

Metal Building Contractors and Erectors Association (MBCEA): A trade organization that supports the advancement of metal building contractors, erectors, and the metal building industry.

Metal Building Institute (MBI): MBI was established to provide educational and training programs for metal building contractors, erectors, and students in construction. Members are manufacturers, contractors, and dealers in two distinct segments of the industry: permanent modular construction (PMC) and relocatable buildings (RB). Associate members are companies supplying building components, services, and financing.

Metal Building Manufacturers Association (MBMA): The association is instrumental in defining and promoting the interests of metal building systems manufacturers. MBMA sponsors research programs to improve the efficiency and quality of metal building systems and to elevate the technology used to produce them.

Metal Construction Association (MCA): An organization of manufacturers and suppliers of metal products. MCA focuses on promoting the use of metal in the building envelope through marketing, education, and action on public policies that affect metal's use.

Metal Framing Manufacturers Association (MFMA): MFMA focuses on the manufacture of ferrous and nonferrous metal framing (continuous slot metal channel systems) that consists of channels with in-turned lips and associated hardware for fastening to the channels at random points.

Metal Powder Industries Federation (MPIF): An association formed by the powder metallurgy (PM) industry to advance the interests of the metal powder–producing and –consuming industries.

Metal Roofing Alliance (MRA): MRA was founded to educate consumers about metal roofing. Membership includes paint companies, material suppliers, industry publications, and others.

Metals Service Center Institute (MSCI): A trade association that serves the industrial metals industry. MSCI provides data and education for operational efficiency, promotes industry advocacy, and creates a marketplace for efficient transactions, debate, discussion, and learning.

Mineral Insulation Manufacturers Association (MIMA): A source of information and advice on rock and glass mineral wool. MIMA promotes the benefits of mineral wool insulation and the contribution it makes to the energy efficiency of buildings and the comfort of their occupants.

Modular Building Institute: A trade organization that strives to expand the use of offsite. It provides outreach and education to the construction community and customers.

Moulding and Millwork Producers Association (MMPA): A trade association whose goals are to promote quality products, develop sources of supply, promote optimum use of raw materials, standardize products, and increase the domestic and foreign usage of molding and millwork products.

MSR Lumber Producers Council: A not-for profit corporation of the State of Washington. The council represents the interests of machine stress–rated lumber producers in the manufacturing, marketing, promotion, utilization, and technical aspects of machine stress rated lumber.

National Air Duct Cleaners Association (NADCA): An association of companies engaged in the cleaning of HVAC systems. It promotes source removal as the only acceptable method of cleaning and establishes industry standards for the association. NADCA also refers to itself as the **HVAC Inspection, Maintenance and Restoration Association**.

National Air Filtration Association (NAFA): A trade association whose members are from air filter and component manufacturers, sales and service companies, and HVAC and indoor air quality professionals.

National Alarm Association of America (NAAA): A trade association that serves as a forum for alarm dealers and as a filter and provider of training programs and manuals for the education of installers, service personnel and system designers.

National American Wholesale Lumber Association (NAWLA): NAWLA members include every aspect of the lumber industry from planting seedlings to selling building materials and wood. NAWLA is an advocate for wood's role in a green economy and a healthy planet.

National Asphalt Pavement Association (NAPA): A trade association that represents asphalt pavement material producers

and paving contractors. NAPA supports an active research program and provides technical, educational, and marketing materials and information to its members, users, and specifiers of paving materials.

National Association of Architectural Metal Manufacturers (NAAMM): The association represents architectural metal products for building construction. NAAMM currently has six operating divisions: Architectural Metal Products (AMP); Detention Equipment Manufacturers Association (DEMA); Expanded Metal Lath Association (EMLA); Expanded Metal Manufacturers Association (EMMA); Hollow Metal Manufacturers Association (HMMA); and, Metal Bar Grating (MBG).

National Association of Catastrophe Adjusters, Inc. (NACA): The association is focused on catastrophe insurance adjusting for its members through education, shared resources, and technology.

National Association of Electrical Distributors (NAED): An organization that serves the electrical distribution channel. NAED provides its members with tools, information, and assistance to help them financially and improve the electrical distribution channel.

National Association of Elevator Contractors (NAEC): A trade association that serves the interests of independent elevator contractors and suppliers of products and services. NAEC promotes safe and reliable elevator, escalator, and short-range transportation and promotes in the management of member companies.

National Association of Elevator Safety Authorities (NAESA): An organization that *fosters and assists in the promulgation of a standard safety code for elevators and related equipment.*

National Association of Home Builders (NAHB): A trade association that promotes housing as a national priority. NAHB's various groups analyze policy issues, work toward improving the housing finance systems, analyze and forecast and consumer trends, and, generally, are involved in all aspects of the housing industry.

National Association of Mold Remediators and Inspectors (NAMRI): NAMRI is a trade organization for mold inspectors and the remediation industry in general. It establishes standards of practice and promotes an ethical code of conduct for the industry. NAMRI offers information for real estate professionals and homebuyers.

National Association for Surface Finishers (NASF): NASF represents the surface coatings industry. NASF advances an environmentally and economically sustainable future for the finishing industry and promotes the role of surface technology in the global manufacturing value chain. The **American Electroplating & Surface Finishing Foundation (AESF)** is part of the NASF. It focuses exclusively on technical, educational, and research programs.

National Association of Pipe Coating Applicators (NAPCA): The association represents plant-applied pipe coating companies worldwide and promotes standardized protective coating practices. NAPCA includes as associate and international associate members firms that service or have a common industry interest in the pipe coating industry.

National Association of Reinforcing Steel Contractors (NARSC): NARSC furthers the interests of reinforcing steel and post-tensioning contractors throughout the United States and Canada. NARSC is in partnership with the **International Association of Bridge, Structural, Ornamental and Reinforcing Iron Workers** and its local unions.

National Association of Sewer Service Companies (NASSCO): The association researches, evaluates, and develops new methods to train and educate its members about the importance of properly rehabilitated underground utilities.

National Association of State Fire Marshals (NASFM): NASFM comprises many senior fire officials in the United States. State Fire Marshals' responsibilities vary from state to state, but Marshals tend to be responsible for fire safety code adoption and enforcement, fire and arson investigation, fire incident data reporting and analysis, public education and advising Governors and State Legislatures on fire protection. Some state Fire Marshals are responsible for fire fighter training, hazardous materials incident responses, wild land fires, and the regulation of natural gas and other pipelines. They use wildland without a space.

National Association of Vertical Transportation Professionals (NAVIP): NAVIP is a trade organization that represents products in the areas of industrial and process equipment and conveying systems.

National Association of Waterproofing and Structural Repair Contractors (NAWSRC): A professional trade association that serves the public and waterproofing, structure, and foundation repair industries.

National Bureau of Standards (NBS): The organization was founded in 1901. Its function is to establish and maintain standards for units of measurements.

National Clay Pipe Institute (NCPI): The institute does research and development of clay pipe technology. NCPI provides assistance in design, training, and evaluation of systems. Forensic analysis is performed when necessary.

National Coil Coating Association (NCCA): A trade organization that is dedicated to the growth of coil-coated products. NCCA's member companies provide the coil coating service and are leading manufacturers and suppliers of metal, coatings, chemicals, and equipment.

National Collegiate Athletic Association (NCAA): NCAA is made up of three membership classifications: Divisions I, II, and III. Each division creates its own rules governing personnel, amateurism, recruiting, eligibility, benefits, financial aid, and playing and practice seasons. NCAA rules set the criteria for college/university athletic fields.

National Concrete Masonry Association (NCMA): A national trade association that represents the concrete masonry industry. NCMA is involved in technical, research, marketing, government relations, and communications activities. NCMA offers technical services and design aids through publications, computer programs, slide presentations, and technical training.

National Corrugated Steel Pipe Association (NCSPA): NCSPA promotes public policy relating to the use of corrugated steel drainage structures. The association collects and distributes technical information, assists in the formulation of specifications and designs, encourages greater knowledge of corrugated steel pipe's benefits and uses among college engineering students, and conducts seminars about the product and its application among designers.

National Council of Acoustical Consultants (NCAC): An international organization that supports the acoustical profession. NCAC comprises professional firms that specialize in acoustical consulting. To qualify for membership, the firm's principals who practice acoustical consulting must be full members of either the **Acoustical Society of America (ASA)** or **Institute of Noise Control Engineering (INCE)**.

National Council of Examiners for Engineering and Surveying (NCEES): The council develops, administers, and scores the examinations used for engineering and surveying licensure in the United States.

National Council on Qualifications for the Lighting Professions (NCQLP): An organization that serves and protects the public through lighting practice. NCQLP establishes the education, experience, and examination requirements for baseline certification across the lighting professions.

National Council on Radiation Protection and Measurements (NCRP): Chartered by the U.S. Congress as the National Council on Radiation Protection and Measurements, NCRP strives to prevent the occurrence of clinically significant radiation-induced deterministic effects of radiation and limit the risk of stochastic effects in exposed persons to an amount that is acceptable in relation to the benefits to the individual and to society.

National Demolition Association (NDA): A trade association that offers demolition-related services and products. It provides information on equipment and services and on regulatory and safety matters.

The National Earthquake Hazards Reduction Program (NEHRP): Congress established NEHRP in 1977, directing that four federal agencies coordinate their complementary activities to implement and maintain the program. These agencies are **FEMA**, the **National Institute of Standards and Technology**, the **National Science Foundation**, and the **U.S. Geological Survey**. NEHRP leads the federal government's efforts to reduce the fatalities, injuries, and property losses caused by earthquakes.

National Electrical Contractors Association (NECA): The association represents electrical contractors from firms of all sizes performing a range of services. Most NECA contractors qualify as small businesses; however, many large, multinational companies are also members of the association.

National Electrical Manufacturers Association (NEMA): An association of electrical equipment manufacturers, its member companies manufacture products such as power transmission and distribution equipment, lighting systems, factory automation and control systems, and medical diagnostic imaging systems.

National Elevator Industry, Inc. (NEII®): A national trade association of the building transportation industry, NEII promotes safety in building transportation, promotes laws and regulations that permit the introduction of safe, innovative technology, and endorses adoption of current model codes.

National Environmental Balancing Bureau (NEBB): Members perform testing, adjusting, and balancing of heating, ventilating, and air-conditioning systems, commission and retro-commission building systems, execute sound and vibration testing, building envelope testing, test and certify laboratory fume hoods, and electronic and biological clean rooms.

National Fenestration Rating Council (NFRC): An organization that administers a uniform, independent rating and labeling system for the energy performance of windows, doors, skylights, and attachment products, NFRC is an **American National Standard Institute (ANSI)**–accredited standards developer (ASD) that develops and administers comparative energy and related rating programs for fenestration products.

National Fire Protection Association (NFPA): An international organization that is dedicated to reducing the burden of fire on peoples' quality of life by proposing codes and standards, research, and education on fire related issues.

National Fire Sprinkler Association (NFSA): NFSA strives to protect lives and property from fire by promoting the widespread acceptance of the fire sprinkler concept. NFSA also provides engineering and training to its members.

National Fireplace Institute® (NFI): A professional certification division of the **Hearth, Patio & Barbecue Education Foundation (HPBEF)** it is an educational organization for the hearth industry.

National Floor Safety Institute (NFSI): A not-for-profit organization whose mission is to aid in the prevention of slips, trips-and-falls through education, research, and standards development.

National Frame Building Association (NFBA): A trade association that promotes the interests of the post-frame construction industry and its members professionals throughout the United States. NFBA's members are primarily post-frame builders, suppliers, manufacturers, building material dealers, code and design professionals, and structural engineers.

National Glass Association (NGA): A trade association that serves the architectural glass, automotive glass, and window and door industries. NGA provides education and training programs that pertain to technical skills, management practices, and quality workmanship.

National Ground Water Association (NGWA): A trade organization that is composed of U.S. and international groundwater contractors, scientists and engineers, equipment manufacturers, and suppliers. NGWA provide guidance to members, government representatives, and the public for sound scientific, economic, and beneficial development, protection, and management of the world's groundwater resources.

National Guild of Professional Paperhangers, Inc. (NGPP): The guild is dedicated to superior craftsmanship in the hanging of every type of wallpaper, including the hanging of historic wallpapers, scenic murals, digital murals, silk papers, bamboo, grass cloth, English pulp papers, as well as traditional fabric-backed and paper-backed vinyl materials. They use grasscloth without a space.

National Hardwood Lumber Association (NHLA): The association was the creator and is the keeper of the North American hardwood lumber grading rules. NHLA provides technical short courses to on-site company training by an NHLA National Inspector; NHLA conducts an Inspector Training School.

National Home Furnishings Association (NHFA): An organization devoted specifically to the needs and interests of home furnishings retailers.

National Institute of Standards and Technology (NIST): A nonregulatory federal agency within the U.S. Department of Commerce, NIST's mission is to promote U.S. innovation and industrial competitiveness by advancing measurement science, standards, and technology in ways that enhance economic security and improve our quality of life.

National Institute of Steel Detailing (NISD): An international association that advocates, promotes, and serves the interests of the steel detailing industry. NISD membership is offered to steel detailing firms and associated companies and individuals.

National Insulation Association (NIA): A trade association that represents both the merit (open shop) and union contractors, distributors, laminators, fabricators, and manufacturers that provide thermal insulation, insulation accessories, and components to the commercial, mechanical, and industrial markets throughout the nation.

National Kitchen and Bath Association (NKBA): A trade association whose membership includes distributors, retailers, remodelers, manufacturers, fabricators, installers, designers, and other professionals. NKBA's certification program emphasizes continuing education and career development.

National Lighting Bureau (NLB): An organization founded to educate lighting decision-makers about the benefits of High-Benefit Lighting®. Professional societies, trade associations, manufacturers, utilities, and agencies of the federal government sponsor the NLB.

National Lightning Safety Institute (NLSI): The institute advocates a risk management lightning protection strategy. NLSI consults to identify vulnerabilities and to organize defenses and teaches and trains personnel.

National Lumber Grades Authority (NLGA): The authority is responsible for the establishment, issuance, publication, amendment, and interpretation of Canadian lumber grading rules and standards.

National Onsite Wastewater Recycling Association (NOWRA): NOWRA's principal purpose is to educate and serve its members and the public by promoting federal, state, and local policy, improving standards of practice, and advancing public recognition of areas that have no wastewater infrastructure.

National Ornamental and Miscellaneous Metals Association (NOMMA): A trade association of the ornamental and miscellaneous metalworking industry, NOMMA's members fabricate everything from railings and driveway gates to structural and industrial products. NOMMA provides continuing education to its members.

(U.S.) National Park Service (NPS): The National Park Service is a bureau of the U.S. Department of the Interior and is led by a director nominated by the President and confirmed by the U.S. Senate. The National

Register of Historic Places, a comprehensive list of districts, sites, buildings, structures, and objects of national, regional, state, and local significance in American history, architecture, archeology, engineering, and culture is kept by the NPS under the authority of the National Historic Preservation Act of 1966.

National Paint and Coatings Association (NPCA): A trade association that represents paint and coatings manufacturers, raw materials suppliers, and distributors. NPCA acts as an advocate on legislative, regulatory, and judicial issues at the federal, state, and local levels.

National Parking Association (NPA): NPA offers education, networking opportunities, advocacy, products, and services to its members.

National Pest Management Association (NPMA): A trade organization that is committed to the protection of public health, food, and property. It supports its members both technically and businesswise.

National Precast Concrete Association (NPCA): An international trade association that provides technical and production information and education and networking opportunities to its members.

National Program for Playground Safety (NPPS): The program's mission is to help the public create safe and developmentally appropriate play environments for children. NPPS provides research, training, and development of play areas.

National Ready Mixed Concrete Association (NRMCA): The association represents and serves the ready mixed concrete industry through leadership, promotion, education, and partnering; it is an advocate for the industry.

The National Restaurant Association (NRA): The NRA is a foodservice trade association that represents and advocates for foodservice industry interests with state, local, and national policymakers. It has no connection to the National Rifle association.

National Roof Deck Contractors Association (NRDCA): A trade association of contractors, manufacturers, and associates who provide, install, or support the application of engineered composite roof deck substrates in the commercial roof top market. NRDCA develops guidelines, procedures, and educational programs.

National Roofing Contractors Association (NRCA): NRCA provides a forum for roofing contractors, manufacturers, and suppliers of programs and projects that contribute to the continual improvement of the roofing industry.

National Sanitation Foundation (NSF): A nationally recognized testing laboratory that certifies plumbing products to meet the standard to which they were created.

National Slate Association (NSA): NSA develops and disseminates technical information, standards, and educational resources on the materials and methods used in the manufacture, design, and construction of slate roofs and associated flashing systems. NSA's members include roofing contractors, slate quarries and distributors, architects, architectural conservators, roofing consultants, craftspeople, building owners, facilities managers, manufacturers, educators, and others concerned with the manufacture, design, construction, and care of natural slate roofs.

National Society of Professional Engineers (NSPE): In partnership with the state societies, NSPE is an organization of licensed professional engineers (PEs) and engineer interns. NSPE provides education, licensure advocacy, leadership training, multidisciplinary networking, and outreach.

National Stone, Sand and Gravel Association (NSSGA): In 2001, the

National Stone Association and **National Aggregates Association** merged to become the NSSGA. NSSGA represents the crushed stone, sand, and gravel—or construction aggregates—industries.

National Storm Shelter Association (NSSA): The association fosters quality by recognizing and distinguishing the shelter producers (and their products) who meet the association's standard of quality. Members include producers, installers, associate members, professionals, media partners, and corporation and individual sponsors.

National Sunroom Association (NSA): A professional organization whose members include manufacturers, design professionals, and material suppliers and installers. It informs consumers, remodelers, and building officials about sunrooms, patio rooms, solariums, and conservatories, and ensures that products are designed and manufactured to high standards and are safety compliant, energy efficient, and environmentally friendly.

National Systems Contractors Association (NSCA): A trade association that represents the low-voltage industry, including systems contractors/integrators, product manufacturers, consultants, sales representatives, architects, specifying engineers, and other allied professionals.

National Terrazzo and Mosaic Association, Inc. (NTMA®, Inc.): NTMA® Inc. is a trade association that establishes national standards for terrazzo floor and wall systems. NTMA also provides specifications, color plates, and other information to architects and designers. Membership is limited to terrazzo contractors.

National Tile Contractors Association (NTCA): A trade association that serves the tile and stone industry. NTCA includes manufacturers, distributors, contractors, architects, designers, and builders in its programs.

National Tile Roofing Manufacturers Association (NTRMA): NTRMA has changed its name to **Tile Roofing Institute**. TRI trains roofing installers, inspectors, and industry professionals on proper, code-approved methods to installing concrete and clay tile roofs. See **Tile Roofing Institute**.

National Utility Contractors Association (NUCA): A trade association dedicated to the underground utility construction industry. NUCA represents contractors, suppliers, and manufacturers involved in water, sewer, gas, electric, telecommunications, site work, and other segments of the industry.

National Wood Flooring Association (NWFA): A trade association that represents the hardwood flooring industry: manufacturers, distributors, retailers, and installers. NWFA provides training and resources to wood flooring.

National Wood Window and Door Association (NWWDA): The Association sets standards for the residential and commercial window, door, and skylight industry. It is a trade organization that advances these standards among industry members and also provides resources, education, and professional programs for its members.

North American Association of Floor Covering Distributors (NAFCD): NAFCD promotes the wholesale distribution of floor coverings and provides members with resources for enhancing performance as industry suppliers. NAFCD represents its members through involvement and by providing education through its development programs and conferences.

North American Association of Food Equipment Manufacturers (NAFEM): A trade association of foodservice equipment and supplies manufacturers. NAFEM offers educational opportunities to its members.

North American Deck and Railing Association (NADRA): A trade association of the deck and railing building industry in North America. NADRA is made up of deck builders, manufacturers, dealers/distributors, wholesalers, retailers, and service providers to the industry.

North American Fiberboard Association (NAFA): A trade organization of manufacturers of cellulosic fiberboard products that are used for residential and commercial construction, commercial products, and packaging.

North American Insulation Manufacturers Association (NAIMA): Member companies manufacture fiber glass, rock wool, and slag wool insulations for residential, commercial, and industrial uses. NAIMA is a resource on energy-efficiency, sustainable performance, and the application and safety of fiberglass, rock wool, and slag wool insulation products.

North American Laminate Flooring Association (NALFA): A trade organization dedicated to the laminate flooring industry. NALFA is an accredited ANSI standards developing organization and publishes testing and performance criteria.

Northeastern Lumber Manufacturers Association (NELMA): NELMA is the rules writing agency for Eastern White Pine lumber and the grading authority for Eastern Spruce, Balsam Fir, Spruce-Pine-Fir (SPFs) grouping, and commercial eastern softwood lumber species. NELMA is an agency for export wood packaging certification. NELMA does marketing for the wood products industry in the Northeast.

Northwest Wall and Ceiling Bureau (NWCB): An international trade association for the wall and ceiling industry, NWCB's membership consists of subcontractors, manufacturers, suppliers, labor organizations, and other professionals in the wall and ceiling industry in the United States and Canada. NWCB works with architects, code bodies, designers, and construction professionals on the design and application of the industry's products.

Occupational Safety and Health Administration (OSHA): A federally funded agency in the Department of Labor that develops job safety and health standards. The states have parallel organizations, e.g., CAL/OSHA.

Painting and Decorating Contractors of America (PDCA): A trade association of painting and decorating contractors, PDCA offers contractor members education programs, attendance at local networking meetings, and use of PDCA industry standards.

Petroleum Equipment Institute (PEI): A trade association whose members manufacture, distribute, and service petroleum marketing and liquid-handling equipment. Members include manufacturers, sellers, and installers of equipment used in service stations, terminals, bulk plants, fuel, oil and gasoline delivery, and similar petroleum marketing operations. No s afte equipment.

Pile Driving Contractors Association (PDCA): The association advocates the use of driven piles for deep foundations and earth retention systems. PDCA promotes the use of driven pile solutions, supports educational programs for engineers and contractors, and encourages and supports research.

Pipe Fabrication Institute® (PFI): PFI promotes standards of excellence in the pipe fabrication industry. PFI initiates research and studies, proposes and maintains standards and technical bulletins, and organizes meetings and other technical exchanges within the industry.

Planet Professional Landcare Network (Planet): A national trade association that represents landscape industry professionals. PLANET was created when the Associated Landscape Contractors of America (ALCA) and the Professional Lawn Care Association of America (PLCAA) merged to create a single national trade association of lawn care and landscape professionals. As they use it.

Plastic Lumber Trade Association (PLTA): A trade association that promotes standardized testing, standards of quality, recycling of plastics, and generally works in support of the plastic lumber industry.

Plastic Pipe and Fittings Association (PPFA): A trade association that promotes and defends plastic piping systems governed by construction codes. PPFA provides users with information to design, specify, and install plastic piping systems. PPFA promotes an understanding of the thermoplastic piping products as they pertain to the environment.

Plastic Pipe Institute (PPI): A trade association that represents all segments of the plastics piping industry. PPI promotes the use of plastics piping for water and gas distribution, sewer and wastewater, oil and gas production, industrial and mining uses, power and communications, duct, and irrigation.

Plumbing and Drainage Institute (PDI): PDI is an association of manufacturers of engineered plumbing products. PDI's members and licensees make products such as: floor drains, roof drains, sanitary floor drains, cleanouts, water hammer arresters, backwater valves, grease interceptors, fixture supports, and other drainage specialties.

Plumbing, Heating, Cooling Contractors Association-National Association (PHCC-National Association): A trade organization for plumbing, heating, and cooling professionals. PHCC promotes advancement, education, and training.

Plumbing Manufacturers International (PMI): A trade association of manufacturers of plumbing industry products such as potable water supply system components,

fixture fittings, waste fixture fittings, fixtures, flushing devices, sanitary drainage system components, and plumbing appliances.

Polyisocyanurate Insulation Manufacturers Association (PIMA): A national trade association that represents polyiso insulation (a widely used insulation product) manufacturers and suppliers to the polyiso industry.

Porcelain Enamel Institute, Inc. (PEI): A trade organization that is dedicated to advancing the porcelain-enameling plants and suppliers of porcelain-enameling materials and equipment.

Portland Cement Association (PCA): PCA represents cement companies in the United States and Canada. The association conducts programs of market development, education, research, technical services, and government affairs on behalf of its members.

Post-Tensioning Institute (PTI): The institute is dedicated to expanding post-tensioning applications through marketing, education, research, teamwork, and code development. PTI advances the quality, safety, efficiency, profitability, and use of post-tensioning systems. Members include post-tensioning materials fabricators, manufacturers of prestressing materials, and companies supplying materials, services, and equipment used in post-tensioned construction.

Powder Actuated Tool Manufacturers' Institute, Inc. (PATMI): An association of manufacturers of powder-actuated fastening systems. PATMI stresses training, certification, and safety awareness. (Powder-actuated fasteners are used to bond various construction materials together, such as wood and concrete or steel and concrete.)

Powder Coating Institute (PCI): A trade organization that represents the North American powder-coating industry, promotes powder-coating technology, and communicates the benefits of powder coating to manufacturers, consumers, and government.

Power and Communication Contractors Association (PCCA): A national trade association for companies that construct electric power facilities, including transmission and distribution lines and substations and telephone, fiber optic, and cable television systems. Other areas of members' business activities include directional drilling, local area and premises wiring, water and sewer utilities, and gas and oil pipelines.

Precast/Prestressed Concrete Institute (PCI): PCI fosters understanding and use of precast and prestressed concrete. Prestressed is correct.

Professional Awning Manufacturers Association of the Industrial Fabrics Association International (PAMA): An international trade association that is committed to supporting the awning industry. Membership is open to companies that are a member of **Industrial Fabrics Association International (IFAI)** and manufacture and/or supply material to the awning industry.

Professional Grounds Management Society (PGMS): Comprising professional grounds managers and other people interested in the grounds management industry, PGMS promotes the dissemination of educational materials and information relevant to the execution of grounds management functions.

Professional Women in Construction (PWC): Originally the National Association of Professional Women in Construction, PWC is an organization that is committed to advancing professional, entrepreneurial, and managerial opportunities for women and other "non-traditional" populations in construction and related industries.

Quality Assurance Association (QAA): An organization of professionals from wholesalers, retailers, manufacturers, laboratories, government, suppliers, and others.

Rack Manufacturers Institute (RMI): An association that advances the standards, quality, and safety of industrial steel storage rack systems.

Radiant Professionals Alliance (RPA): The alliance promotes radiant heating on behalf of its members: contractors and dealers, manufacturers, distributers, designers, and others with an interest in the industry.

Reflective Insulation Manufacturers Association International (RIMA-I): A trade association that represents the reflective insulation, radiant barrier, and low-e reflective coatings industries.

Reflective Roof Coatings Institute (RRCI): A trade organization whose members are coatings manufacturers, raw materials suppliers, applicators, and industry consultants.

Research Council on Structural Connections (RCSC): A nonprofit organization that comprises leading experts in the fields of structural steel connection design, engineering, fabrication, erection, and bolting.

Resilient Floor Covering Institute (RFCI): An industry trade association of resilient flooring manufacturers and suppliers of raw materials, additives, and sundry flooring products for the North American market.

Restoration Industry Association (RIA): A trade association for cleaning and restoration professionals. RIA provides leadership and promotes best practices through advocacy, standards, and professional qualifications for the restoration industry.

Retail Contractors Association (RCA): A national organization of retail contractors who have united to provide a solid foundation of ethics, quality, and professionalism within the retail construction industry.

Roof Coatings Manufacturers Association (RCMA): An association of roof coating manufacturers and affiliates that advances, promotes, and expands the international market for roof coatings through education, technical advancement, and advocacy of industry issues.

Roof Consultants Institute, Inc. (RCI, Inc.): An international association of professional consultants, architects, and engineers who specialize in the specification and design of roofing, waterproofing, and exterior wall systems.

Roofing Industry Educational Association (RIEI): A roofing industry education resource. In 2000, RIEI merged with the **National Roofing Contractors Association (NRCA)**, and together they provide a variety of courses and training seminars.

Rubber Manufacturers Association (RMA): A national trade association for tire manufacturers that make tires in the United States.

Scaffolding, Shoring and Forming Institute (SSFI): A trade association of manufacturers of scaffolding, suspended scaffolding, shoring, forming, planks, platforms, and related components. The institute focuses on the technical aspects and safe use of scope products.

Safety Glazing Certification Council (SGCC): A not-for-profit corporation comprising manufacturers of safety glazing products, building code officials, and others concerned with public safety. SGCC maintains a program that provides for independent, third-party certification of safety glazing materials.

Scientific Certification Systems (SCS): A third-party provider of certification, auditing, and testing services of forest management operations and wood product manufacturers; SCS is accredited by the Forest Stewardship Council (FSC). SCS evaluates forests according to the FSC Principles and Criteria for Forest Stewardship.

Scientific Equipment and Furniture Association (SEFA): A trade organization of lab designers and manufacturers of laboratory furniture. Members include companies whose work is principally in this industry.

Screen Manufacturers Association (SMA): A trade organization that represents the window and door screen industry. SMA participates in creating standards that meet various governmental and code entity requirements.

Sealant Waterproofing and Restoration Institute (SWRInstitute): An international trade association that represents the commercial sealant, waterproofing, and restoration construction industry. SWRInstitute's members include contractors, manufacturers, and design professionals in the industry.

Security Industry Association (SIA): A trade organization that advocates pro-industry policies and legislation, produces market research, creates open industry standards, provides education and training, and opens global market opportunities.

Sheet Metal and Air Conditioning Contractors National Association (SMACNA): An international trade association that promotes quality and excellence in sheet metal and air conditioning technology and construction.

Siding and Window Dealers Association of Canada (SAWDAC): Members are dealers and contractors of products and installations. Members must commit to SAWDAC's code of ethics and sign a 5-year workmanship guarantee statement.

Single Ply Roofing Industry (SPRI): SPRI represents sheet membrane and related component suppliers in the commercial roofing industry. SPRI serves as a commercial roofing components and system information resource for building owners, architects, engineers, designers, contractors, and maintenance personnel.

Slag Cement Association (SCA): A trade organization that represents companies that produce and ship slag cement (ground granulated blast furnace slag) in the United States.

Society of American Military Engineers (SAME): A professional military engineering association in the United States. SAME unites architecture, engineering, construction (A/E/C), facility management, environmental firms, and individuals in the public and private sectors to prepare for and overcome natural and manmade disasters and to improve security at home and abroad.

Society of Automotive Engineers International (SAE): An organization comprising engineers and related technical experts in the aerospace, automotive, and commercial-vehicle industries. SAE's core competencies are life-long learning and voluntary consensus standards development. The society provides standards, e.g., a thread size used on nuts and bolts but not pipe connections.

Society of Fire Protection Engineers (SFPE): A professional society that represents people who practice fire protection engineering. SFPE advances the science and practice of fire protection engineering and its allied fields, maintains a high ethical standard among its members and fosters fire protection engineering education. SFPE supports the development of the annual Professional Engineer licensing examination in fire protection and the grading of those examinations under the auspices of the National Council of Examiners for Engineering and Surveying.

Society of Glass and Ceramic Decorated Products (SGCDpro): The society is dedicated to the interests of the decorating, manufacturing, and marketing of glass and ceramics and associated businesses. SGCDpro provides information about business opportunities and technical, educational, and regulatory information to its members. SGCDpro promotes the use of socially and environmentally responsible business practices.

Society of the Plastics Industry (SPI): SPI promotes business development, fosters the sustainable growth of plastics in the global marketplace, provides industry representation in the public policy arena, and communicates the industry's contributions to the society and the benefits of its products.

Society for Protective Coatings (SSPC): A professional technical society whose primary objective is to improve the technology and practice of prolonging the life of steel and concrete structures through the use of protective coatings. SSPC was originally **Steel Structures Painting Council**.

Society of Wood Science and Technology (SWST): The society develops and maintains knowledge that is specific to the science and technology of wood and other lignocellulosic materials; encourages the use of this knowledge, promotes policies and procedures that are aimed at wise and responsible use of wood and other lignocellulosic materials (any of several closely related substances constituting the essential part of woody cell walls of plants and consisting of cellulose intimately associated with lignin); assures high standards for members; fosters educational programs at all levels of wood science and other lignocellulosic materials and their technologies and furthers the quality of such programs; and represents the wood science and technology profession in public policy development.

Soil and Water Conservation Society (SWCS): A scientific and educational organization that serves as an advocate for conservation professionals and for science-based conservation practice, programs, and policy. SWCS members include researchers, administrators, planners, policymakers, technical advisors, teachers, students, farmers, and ranchers.

**Solar Energy Industries Association®
(SEIA):** A trade organization of the solar energy industry, SEIA's member companies research, manufacture, distribute, finance, and build solar projects domestically and abroad.

**Solar Rating and Certification™
(SRCC™):** An organization whose primary purpose is to provide authoritative performance ratings, certifications, and standards for solar thermal products.

Solar and Sustainable Energy Society of Canada Inc. (SESCI): The society advances the use and awareness of solar and sustainable energy in Canada.

Southern Building Code Congress: An active participant in the International Code Council, SBCCI helps to develop and maintain the ICC model building codes. It also continues to provide prints of its original codes that were published prior to merging with the ICC.

Specialty Steel Institute of North America (SSINA): The institute is a voluntary trade association that represents producers of specialty steel in North America. Its members produce a variety of products including bar, rod, wire, angles, plate, sheet and strip, in stainless steel and other specialty steels.

Southern Coast Air Quality Management District (SCAQMD): The air pollution agency responsible mainly for regulating stationary sources of air pollution for most of Los Angeles, San Bernardino, Riverside County, and all of Orange County. SCAQMD develops, adopts, and implements an Air Quality Management Plan for bringing the area into compliance with the clean air

standards established by national and state governmental legislation. It also is a model for air quality management throughout the country.

Southern Pine Inspection Bureau (SPIB®): SPIB® comprises family-owned and publicly traded companies that place SPIB®'s logo on their products.

Specialty Steel Industry of North America (SSINA): A trade association that represents producers of specialty steel in North America. SSINA members produce products including bar, rod, wire, angles, plate, sheet, and strip in stainless steel and other specialty steels.

Spiral Duct Manufacturers Association (SPIDA): A trade organization that promotes the use of round duct, spiral duct (spiral pipe), and flat oval duct; supports testing and research of round duct and spiral pipe; and provides manufacturers with specialized information.

Spray Polyurethane Foam Alliance (SPFA): A trade organization that also serves as an educational and technical resource for the spray polyurethane foam industry.

Stairway Manufacturers' Association (SMA): A trade association that serves stairway manufacturers, SMA provides code officials, design professionals, builders, and the stair industry and community schools with technical expertise through publications, seminars, and direct classroom experiences.

Steel Deck Institute (SDI): SDI keeps designers and constructors up-to-date on deck design and construction and to provides information about the SDI and the member companies.

Steel Door Institute (SDI): In conjunction with other testing laboratories such as UL, SDI tests steel doors and frames for strength, quality, consistency, security, weather and fire resistance, wear and tear, and longevity. SDI also works alongside industry associations representing related products to ensure compatibility with products used in conjunction with steel doors and frames.

Steel Erectors Association of America (SEAA): An organization that sets uniform standards among the many steel erectors and helps promote safety in the erection industry.

Steel Framing Alliance (SFI): SFI works to expand market share in the commercial and residential construction markets, with an emphasis on growth potential of structural cold-formed steel (CFS) framing in the mid-rise sector.

Steel Joist Institute (SJI): An organization of active joist manufacturers that cooperates with business and government agencies to establish steel joist standards. The institute does continuing research of industry products to maintain the integrity of these products.

Steel Manufacturers Association (SMA): Most of SMA's members are electric arc furnace steel producers, or "minimills," that use a feedstock almost entirely composed of recycled steel scrap to make new steel. SMA's associate member companies provide equipment, supplies, and services to steel companies.

Steel Recycling Institute (SRI): An industry association that promotes and sustains the recycling of all steel products. SRI educates the solid waste industry, government, business, and ultimately the consumer about the benefits of steel's infinite recycling cycle.

Steel Stud Manufacturers Association (SSMA): A trade organization of the steel framing manufacturing industry, SSMA represents member firms engaged in the manufacture, marketing, and sale of cold-formed steel framing. Members include contractors, distributors, design professionals, code officials, and standards organizations.

Steel Tube Institute (STI): A trade organization that promotes and markets steel tubing, STI's active membership consists of producers of steel tube and pipe and its associate membership consists of companies that supply raw materials, equipment, and support services.

Steel Window Institute (SWI): A trade organization of manufacturers of windows made from either solid or formed sections of steel, and such related products as casings, trim, mechanical operators, screens, and moldings when manufactured and sold by members of the industry for use in conjunction with windows. SWI provides the public with general and technical information concerning the industry's products.

Structural Building Components Association (SBCA): An international trade association that represents manufacturers of structural building components. Membership also includes truss plate and original equipment manufacturers, computer engineering and other service companies, lumber mills, inspection bureaus, lumber brokers and distributors, builders and professional individuals in the fields of engineering, marketing, and management.

Structural Insulated Panel Association (SIPA): A trade association that represents manufacturers, suppliers, dealer/distributors, design professionals, and builders committed to providing quality structural insulated panels for all segments of the construction industry.

Structural Stability Research Council (SSRC): SSRC offers guidance to specification writers and practicing engineers by developing both simplified and refined calculation procedures for the solution of stability problems and assessing the limitations of these procedures. SSRC is made up of representatives from government agencies, international organizations, private corporations, educational institutions, representatives of consulting firms, members-at-large selected from universities and design offices, and corresponding members from various countries.

Stucco Manufacturers Association (SMA): A trade association that is comprises manufacturers of stucco in North America and their related suppliers.

Submersible Wastewater Pump Association (SWPA): A national trade association that represents and serves the manufacturers of submersible pumps for municipal and industrial wastewater applications. Regular members are manufacturers of submersible wastewater pumps for municipal and industrial. Component members are manufacturers of component parts and accessory products for submersible pumps and pumping systems. Associate members are nonmanufacturers providing services related to industry products and who provide services to the users of industry products.

Surface Mount Technology Association (SMTA): An international trade association that works to develop solutions in electronic assembly technologies, including microsystems, emerging technologies, and related business operations.

Sustainable Forestry Initiative (SFI): An independent organization that promotes responsible forest management.

Telecommunications Industry Association (TIA): TIA is accredited by the **American National Standards Institute (ANSI)** to develop voluntary, consensus-based industry standards for a variety of information and communications technology (ICT) segments. TIA operates twelve engineering committees, that develop guidelines for private radio equipment, cellular towers, data terminals, satellites, telephone terminal equipment, accessibility, voice over internet protocol (VoIP) equipment, structured cabling, data centers, mobile device communications, multimedia multicast, vehicular telematics, healthcare ICT, smart device communications, smart utility mesh networks, and sustainable/environmental communications technologies.

Terrazzo Tile and Marble Association of Canada (TTMAC): A trade organization whose overall objective is to raise the profile of the industry within the marketplace and the respective standards in order to achieve that goal.

Testing, Adjusting and Balancing Bureau (TABB): The **Sheet Metal and Air Conditioning Contractors National Association (SMACNA)** endorses TABB's procedures that include a strict code of conduct for technicians performing hands on TAB work; TABB certification is encouraged for inclusion in project specifications.

The Masonry Society (TMS): Members are design engineers, architects, builders, researchers, educators, building officials, material suppliers, manufacturers, and other interested people. TMS gathers and disseminates technical information through its committees, publications, codes and standards, slide sets, videotapes, computer software, newsletter, refereed journal, educational programs, professors' workshop, scholarships, certification programs, disaster investigation team, and conferences.

Tile Contractors Association of America (TCAA): TCAA is a trade organization that promotes the tile industry. TCAA stresses professionalism, reliability, skilled craftsmanship, and technical performance in the industry. TCAA is in partnership with **International Masonry Institute (IMI)** to develop the "Trowel of Excellence."

Tile Council of North America, Inc. (TCA): A trade association that represents North American manufacturers of ceramic tile, tile installation materials, tile equipment, raw materials, and other tile-related products.

Tile Roofing Institute (TRI): Formerly named the **National Tile Roofing**

Manufacturers Association (NTRMA). Today, TRI trains roofing installers, inspectors, and industry professionals on proper, code-approved methods to installing concrete and clay tile roofs. TRI is dedicated to growing the tile roofing market through technical expertise, training, and building awareness for the many benefits of tile.

Tilt-Up Concrete Association (TCA): A trade association that strives to expand and improve the use of Tilt-Up as a building system. TCA provides education and resources that enhance quality and performance.

Tree Care Industry Association (TCIA): A trade association of commercial tree care firms and affiliated companies. TCIA develops safety and education programs, standards of tree care practice, and management information for arboriculture firms around the world.

Tropical Forest Foundation (TFF): An international educational institution committed to advancing environmental stewardship, economic prosperity, and social responsibility through sustainable forest management (SFM).

Truss Plate Institute (TPI): A trade organization of the plate truss industry. TPI establishes methods of design and construction for trusses in accordance with the **American National Standards Institute's** accredited consensus procedures for the coordination and development of American National Standards. TPI also provides a quality assurance inspection program and contributes its expertise in other technical areas.

Turfgrass Producers International (TPI): A worldwide association committed to the advancement of the turfgrass sod industry. TPI provides education to members, product users, and various green industry and government entities. TPI encourages the use of turfgrass sod worldwide.

Underwriters Laboratories Inc. (UL): UL provides safety-related certification, validation, testing, inspection, auditing, advising, and training services to a wide range of clients, including manufacturers, retailers, policymakers, regulators, service companies, and consumers. A UL label is often displayed on packaging. For example, a UL label on the packaging of shingles indicates the level of fire and wind resistance of asphalt roofing.

Underwriters Laboratories of Canada (ULC): The Canadian equivalent of UL.

Uni-Bell PVC Pipe Association: An association of PVC pipe and fittings' manufacturers that was created to promote and provide technical support in the use of PVC pipe and fittings. Uni-bell writes installation guidelines that are used as a reference regarding installation failures.

United Lighting Protection Association (ULPA): A trade association of lighting protection manufacturers, engineers, contractors, and technicians.

United States Geological Survey: A science organization that provides information on the health of our ecosystems and environment, the natural hazards that threaten us, the natural resources we rely on, the impacts of climate and land-use change, and the core science systems that help us provide timely, relevant, and useable information.

United States Green Building Council (USGBC): USGBC is best known for its development of the Leadership in Energy and Environmental Design (LEED) green building rating systems and its annual Greenbuild International Conference and Expo. The council works toward its mission of market transformation through its LEED program, educational offerings, a nationwide network of chapters and affiliates, the annual Greenbuild International Conference &

Expo, and advocacy in support of green buildings and communities.

United States Sign Council (USSC): An educational resource for the sign industry, USSC is open to any person or firm concerned with the advancement of the sign industry.

Valve Manufacturers Association of America (VMA): VMA represents the interests of North American manufacturers of valves and actuators.

Vibration Isolation and Seismic Control Manufacturers Association (VISCMA): A professional organization consisting of partnerships, companies, and corporations that engage in the seismic restraint, vibration isolation, or noise isolation industry.

Vinyl Institute: A trade association that represents manufacturers of vinyl, vinyl chloride monomer, vinyl additives and modifiers, and vinyl packaging materials.

Vinyl Siding Institute, Inc. (VSI): A trade association for manufacturers of vinyl and other polymeric siding and suppliers to the industry. VSI is the sponsor of the VSI Product Certification Program and the VSI Certified Installer Program.

Warnock Hersey: A Canadian lab that is universally recognized in Canada. WH certifies products to CSA standards.

Wallcovering Association (WA): A trade association that represents wallcoverings manufacturers, distributors, and suppliers to the industry.

Walnut Council: The Walnut Council promotes sustainable forest management and utilization of American black walnut and other high-quality fine hardwoods. Its purpose is to assist in the technical transfer of forest research to field applications, help build and maintain better markets for wood products and nut crops, and to promote sustainable forest management, conservation, reforestation, and utilization of American black walnut (Juglansnigra) and other high-quality fine hardwoods.

Water and Sewer Distributors of America (WASDA): Comprising distributors and manufacturers of waterworks and wastewater products, WASDA promotes the waterworks/wastewater products distribution industry.

Water Environmental Federation (WEF): WEF is similar to **AWWA**. Its members are people associated with water waste including all types of materials.

Water Quality Association (WQA): An international trade association that represents the residential, commercial, and industrial water treatment industry. WQA provides information to the industry, educators, and professionals. WQA has a laboratory for product testing and is a communicator to the public.

West Coast Lumber Inspection Bureau (WCLIB): A service corporation for the benefit and protection of buyers, sellers, and consumers of softwood lumber. WCLIB's primary objective is the development and maintenance of uniform lumber standards.

Western Hardwood Association (WHA): A trade organization that promotes and markets western hardwoods. WHA provides education for stakeholders on sustainable and environmentally responsible resource management.

Western Red Cedar Lumber Association (WRCLA): A Vancouver-based association that represents producers of Western Red Cedar lumber products in Washington, Oregon, and British Columbia (Canada). WRCLA operates customer service programs throughout the United States and Canada to support its members' cedar products with information, education, and quality standards.

Western States Clay Products Association (WSCPA): A trade association of brick manufacturers in the Western United States, WSCPA develops technical information for enhancing quality use of clay brick construction with special attention directed to the seismic performance of clay brick.

Western States Roofing Contractors Association (WSRCA): Members are roofing contractors and vendors. WSRCA serves both members and consumers alike.

Western Wall & Ceiling Contractors Association (WWCCA): An organization that represents subcontractors and affiliates who have joined to promote the installation of quality construction. Members employ only union-trained craftspeople.

Western Wood Preservers Institute (WWPI): A trade organization that represents the preserved wood products industry throughout western North America. WWPI's primary activity areas are regulatory and market outreach programs.

Western Wood Products Association (WWPA): A trade association that represents softwood lumber manufacturers in the Western United States including Alaska. WWPA's mills produce lumber from Western softwood species, including Douglas Fir, Western Larch, Western Hemlock, True Firs, Engelmann Spruce, Ponderosa Pine, Lodgepole Pine, Sugar Pine, Idaho White Pine, Western/Inland Red Cedar, and Incense Cedar.

Window Coverings Association of America (WCAA): A national trade association whose members are from the retail window coverings industry and its dealers, decorators, designers, and workrooms, WCAA provides members educational opportunities, encourages a code of ethics for fair practices, and works for the betterment of the retail window coverings industry.

Window and Door Manufacturers Association (WDMA): Formerly the

National Wood Window and Door Association, this trade organization has established many standards related to wood window and door products.

Wire Association International, Inc. (WAI, Inc.): A technical society for wire and cable industry professionals, WAI promotes, collects, and disseminates technical, manufacturing, and general business information to the ferrous, nonferrous, electrical, fiber optic, and fastener segments of the wire and cable industry.

Wire Reinforcement Institute (WRI): WRI is a trade organization that provides information to the concrete construction industry on the uses of welded wire reinforcement (WWR) and related products.

Women Contractors Association (WCA): An organization composed of women owners and decision-making executives within the construction industry. WCA provides networking opportunities and information specific to the construction industry and the small business owner.

Wood Component Manufacturers Association (WCMA): The association represents manufacturers of dimension and wood component products that supply components for cabinetry, furniture, architectural millwork, closets, flooring, staircases, building materials, and decorative/specialty wood products made from hardwoods, softwoods, and a variety of engineered wood materials.

Wood Products Manufacturers Association (WPMA): The association provides members with information resources and services. WPMA acts as a clearinghouse for solving problems of mutual concern and assists members in controlling costs.

Woodwork Institute (WI): WI provides standards and quality control programs for the architectural millwork industry.

World Floor Covering Association (WFCA): A source of information on all types of flooring, including carpet, hardwood flooring, laminate floors, ceramic tile, area rugs, natural stone, cork, bamboo, and vinyl flooring.

World Waterpark Association (WWA): A trade organization that provides information on waterpark-business topics, industry trends, and other matters that are relevant to the industry.

Woven Wire Products Association (WWPA): WWPA is a trade organization that represents manufactures of diamond-woven wire mesh products for institutional, industrial, and architectural applications. Membership is open to persons, partnerships, or corporations involved in the manufacturing of diamond-woven wire mesh products and those firms that supply materials and services to the manufactures of diamond-woven wire mesh products.

BIBLIOGRAPHY

Here listed are most, but not all, of the references that were used in developing the checklists that are in this book. When you must go beyond these references, the resource organizations listed in Appendix B should get you started in finding the specific information you need.

ADA Evaluation. *The Americans with Disability Act.* Memphis, IN: ThyssenKrupp Elevator, 2006.

American Concrete Institute (ACI). ACI 117-90 *Standard Specifications for Tolerances for Concrete Construction and Materials.* Farmington Hills, MI: ACI, 2002.

American Concrete Institute (ACI). ACI 347-04, *Guide to Formwork for Concrete.* Farmington Hills, MI: ACI, 2004.

American Concrete Institute (ACI). ACI 347, *Recommended Practice for Concrete Formwork.* Farmington Hills, MI: ACI, 2008.

American Concrete Institute (ACI). ACI 301, *Structural Concrete for Buildings.* Farmington Hills, MI: ACI, 2010.

American Concrete Institute (ACI). ACI 318, *Building Code Requirements for Reinforced Concrete.* Farmington Hills, MI: ACI, 2011.

American Society of Civil Engineers. SEI/ASCE 11-99, *Guidelines for Structural Condition Assessment of Existing Buildings.* Reston, VA: ASCE, 2000.

American Society of Civil Engineers. ASCE/SEI 31-03, *Seismic Evaluation of Existing Buildings.* Reston, VA: ASCE, 2003.

Ching, F.D.K. *Building Construction Illustrated.* Hoboken, NJ: John Wiley & Sons, Inc., 2008.

City of Houston. *High Rise Inspection Guide.* Houston, TX: Houston Fire Department.

City of Milpitas. *Commercial Rough & Final Electrical Inspection Checklist.* Milpitas, CA: City of Milpitas, Building and Safety Department.

City of Milpitas. *Plumbing Rough & Final Inspection Checklist.* Milpitas, CA: City of Milpitas, Building and Safety Department.

City of Milpitas. *Residential Window & Door Replacement.* Milpitas, CA: City of Milpitas, Building and Safety Department.

City of New York. *NYC Construction Codes,* 2008.

City of Palo Alto. *Electrical Sub Panel Inspection.* Palo Alto, CA: City of Palo Alto, 2012.

City of Palo Alto. T*ankless Water Heater.* Palo Alto, CA: City of Palo Alto, 2013.

City of Rancho Palos Verdes. *Residential Water Heater Inspection Checklist.* Rancho Palos Verdes, CA: Community Development Department.

City of Wolfeboro. *Oil Burner Inspection Checklist.* Wolfeboro, NH: Fire-Rescue Department. 2002.

Department of Public Health Drinking Water Section. Section 19-13-B51f(e), *Well Casing Extensions*. CT: Regulations of Connecticut State Agencies (RCSA), undated.

Grew, M.G. *Residential Framing Inspections*, Based on 2005 CSBC/2003 IRC. AIA, 2012 Career Development Seminar, Office of Education and Data Management, Department of Construction Services, 2012.

Griffin, C.W. Jr. *Manual of Built-up Roof Systems*. Arlington, TX: McGraw-Hill Book Company, 1970.

Heimer Engineering PC. *The Structure of the Building, 2014*.

Ingersoll Rand, Security Technologies. *Fire Door Assembly Inspection Checklist*. Needham MA: Ingersoll Rand Security Technologies of New England, 2010.

International Association of Certified Home Inspectors. *International Standards of Practice for Inspecting Commercial Properties*. Boulder, CO: InterNACHIcomsop-2014, 2014.

International Building Code Council. *International Building Code*. International Code Council, 2012.

International Building Code Council. *International Residential Code for One and Two-Family Dwellings*. International Code Council, 2012.

International Code Council. *Seattle Building Code Collection Website*. Country Club Hills, IL: International Code Council, 2009.

Laefer, D.F.; J. Gannon; and E. Deely. "Reliability of Crack Detection for Baseline Condition Assessments." *Journal of Infrastructure Systems* 16, no. 2 (May 2010), pp. 129–137.

Lemieux, D.J.; and P.E. Totten. *Building Envelope Design Guide-Wall Systems*. Boston, MA: Wiss, Janney, Elstner Associates, Inc., 2010.

Kaiser, H.H. *The Facilities Audit*. Alexandria, VA: The Association of Higher Education Facilities Officers (APPA), 1993.

Minnesota Department of Labor and Industry, Electrical licensing and Inspection. *Residential Electrical Inspection Checklist*. St Paul, MN: Minnesota Department of Labor and Industry, undated (based on NEC 2011).

My BuildingPermit.Com:

My BuildingPermit.Com. *Residential Building Final*. Washington, DC: My BuildingPermit. Com, January 2005.

My BuildingPermit.Com. *Residential Mechanical Final*. Washington, DC: My BuildingPermit.Com, April 2005.

My BuildingPermit.Com. *Residential Framing*. Washington, DC: My BuildingPermit.Com, July 2005.

My BuildingPermit.Com. *Residential Insulation*. Washington, DC: My BuildingPermit. Com, July 2007.

My BuildingPermit.Com. *Residential Plumbing Rough In*. Washington, DC: My BuildingPermit.Com, April 2011.

National Fire Protection Association. *National Electric Code, NFPA 70*. Quincy, MA: NFPA, 2014.

National Resource Conservation Service, Michigan State. *Michigan Training Module, Concrete Construction Inspection*. MI: USDA, 2009.

Peoria County Department of Planning & Zoning. *Building Inspections Checklist* (covers footings, framing, and final inspections), undated.

Stanford University. *Facilities Design Guidelines*. Stanford, CA: Stanford University, Department website, 2014..

Thiele, T. *Electrical Inspector Checkpoints*. About.Com Electrical, (undated).

University of California, *UCSB Injury and Illness Prevention Program, Developing Your Work-site Specific Safety Inspection Checklist*. Santa Barbara: 2012.

University of Michigan. Engineering and Construction, Occupational Safety &Environmental Health. *Construction Safety Requirements Department of Architecture*, 2010.

U.S. Army. *Inspection and Preventive Maintenance*. In *Technical Manual 5-650, Central Boiler Plants*. Department of Army, 1989.

U.S. Department of the Interior, Bureau of Reclamation. *Guide to Protective Coating Inspection and Maintenance*. Washington, DC: U.S Department of the Interior, 2002.

U.S. Department of the Interior, U.S Fish and Wildlife Service, Division of Engineering. *Construction Inspection Handbook (360 FW 4)*. Washington, DC: U.S Department of the Interior, 2004.

U.S. Department of Labor, OSHA. *Safety and Health Regulations for Construction*, Standards - 29 CFR.

U.S. Green Building Council. *Existing Buildings: Operations & Maintenance, Reference Guide*. 2006 and 2008.

Wilshere, C.J. *Formwork*. London, UK: Thomas Telford, 1989.

INDEX

www.ingramcontent.com/pod-product-compliance
Lightning Source LLC
Chambersburg PA
CBHW082005190326
41458CB00010B/3080